The Systematics Association Special Volume Series 70

Biogeography in a Changing World

T0188143

The Systematics Association Special Volume Series

Series Editor

Alan Warren

Department of Zoology, The Natural History Museum,
Cromwell Road, London SW7 5BD, UK.

The Systematics Association promotes all aspects of systematic biology by organizing conferences and workshops on key themes in systematics, publishing books and awarding modest grants in support of systematics research. Membership of the Association is open to internationally based professionals and amateurs with an interest in any branch of biology including palaeobiology. Members are entitled to attend conferences at discounted rates, to apply for grants and to receive the newsletters and mailed information; they also receive a generous discount on the purchase of all volumes produced by the Association.

The first of the Systematics Association's publications *The New Systematics* (1940) was a classic work edited by its then-president Sir Julian Huxley, that set out the problems facing general biologists in deciding which kinds of data would most effectively progress systematics. Since then, more than 70 volumes have been published, often in rapidly expanding areas of science where a modern synthesis is required.

The *modus operandi* of the Association is to encourage leading researchers to organize symposia that result in a multi-authored volume. In 1997 the Association organized the first of its international Biennial Conferences. This and subsequent Biennial Conferences, which are designed to provide for systematists of all kinds, included themed symposia that resulted in further publications. The Association also publishes volumes that are not specifically linked to meetings and encourages new publications in a broad range of systematics topics.

Anyone wishing to learn more about the Systematics Association and its publications should refer to our website at www.systass.org.

Other Systematics Association publications are listed after the index for this volume.

The Systematics Association Special Volume Series 70

Biogeography in a Changing World

Edited by

Malte C. Ebach

Université Pierre et Marie Curie
Paris, France

Raymond S. Tangney

National Museum Wales
Cardiff, U.K.

CRC Press
Taylor & Francis Group
Boca Raton London New York

CRC Press is an imprint of the
Taylor & Francis Group, an **informa** business

CRC Press
Taylor & Francis Group
6000 Broken Sound Parkway NW, Suite 300
Boca Raton, FL 33487-2742

First issued in paperback 2019

© 2007 by The Systematics Association
CRC Press is an imprint of Taylor & Francis Group, an Informa business

No claim to original U.S. Government works

ISBN 13: 978-0-8493-8038-9 (hbk)
ISBN 13: 978-0-367-38998-7 (pbk)

Library of Congress Cataloging-in-Publication Data

Biogeography in a changing world / [edited by] Malte C. Ebach and Raymond S. Tangney.
    p.cm. – (Systematics Association special volume ; no. 70)
    Includes bibliographical references.
    ISBN 0-8493-8038-3
    1. Biogeography—Congresses. I. Ebach, Malte C. II. Tangney, Raymond S. III. Series.

QH84.B546 2006
578.09—dc22   2006045578

**Visit the Taylor & Francis Web site at
http://www.taylorandfrancis.com**

**and the CRC Press Web site at
http://www.crcpress.com**

Preface

The study of plant and animal distributions has a history as long as biology itself, and throughout its development, biogeography has been closely connected to other biological disciplines. Despite the many reasons for its importance to contemporary biology — not the least of which is its capacity to synthesise data across disciplines — biogeography struggles to find its own identity. It occupies an ambivalent position between that of an adjunct whose status is contingent on other areas of study, such as ecology, evolution, taxonomy, or molecular systematics, and that of an independent discipline with a core of accepted knowledge and methodological principles that guide research. The reciprocity afforded by this marginal position, the source of its capacity for synthesis, provides biogeography with the benefit of interaction with various subjects, but biogeographers are hampered in their attempts to construct a strictly 'biogeographic' approach by the input of approaches and methods from these diverse influences

This volume is in large part based on the papers presented at a symposium titled "What is Biogeography?" held in Cardiff in August 2005, part of the Fifth Biennial Meeting of the Systematics Association hosted by Cardiff University and Amgueddfa Cymru–National Museum Wales.

Our aim in planning the programme for the symposium was to provide a forum for a range of viewpoints in biogeography and the different ways in which it is practised. Our contributors, current leading proponents of differing methods within biogeography, were asked to talk on their particular field and its relevance to the symposium title. By addressing biogeography at this fundamental level, we wished to provide insights into the nature of biogeography and to more clearly define both the differing perspectives within biogeography and the differing methodological approaches. This provides an opportunity to refocus on the way in which some of the major problems in biogeography are being currently addressed, for example, the explanatory power in biogeography of biological processes such as vagility and survivability, the importance of spatial pattern in biogeography, the role of geological reconstructions in biogeographic explanation, as well as the relationship between biogeography and other disciplines. A central issue in historical biogeography is the relation of biological data to geological data and the extent to which biogeography is able to challenge current geological understanding. In this context the question of expanding earth is revisited.

The aim of this volume is to bring together the papers from the meeting to produce a broad-based perspective on the nature of biogeography, offering both historical perspectives based on current understanding and methodological advances, as well as what the future might hold.

We thank all the authors for their contributions and the many reviewers for their helpful comments on drafts of the chapters. We are grateful to the Systematics

Association for sponsoring the symposium and to Amgueddfa Cymru–National Museum Wales and Cardiff University for their generous support in hosting the Biennial Meeting in Cardiff.

Malte C. Ebach and Ray Tangney
Paris and Cardiff

A Pangea That Covers the World

Neal Adams

Among the theories of tectonics that come and go is the tenacious "Growing Earth Theory" most famously promulgated from the 1950s to 1980s by Professor S. Warren Carey from the University of Tasmania. Although the scientific community rejected this theory with the advent of the theory of subduction, this old concept has found new adherents who contend:

1. That the question of whether subduction keeps pace with seafloor spreading or even occurs at all still remains open.
2. With the Big Bang now being attacked from all sides, perhaps the answer to, "How was the universe created?" may be related to, "How can the Earth grow?"

After all, if the sun grows, if the universe grows (rather than explodes), if Jupiter has grown to become a meteorite umbrella to Mars and Earth, if dinosaurs existed 100 mya that appeared much too large to support their weight, if mountains are a 'new' feature on Earth, and so on, is it not incumbent on the scientific community to address the possibility of Earth Expansion and research the same tectonic and biotic links across the Pacific as was done for the Atlantic more than 80 years ago? Is it not incumbent on planetary scientists, who have recently discovered gross tectonic rifting and spreading on Mars, Ganymede, Europa, etc. — but not subduction — to relate these discoveries to the Earth? Is it not incumbent on the scientifi community to explain why all of the continental plates fit neatly together on a much smaller Earth or what happened to the missing Archaen crust? We need an all inclusive theory, based on many different disciplines, that will answer these questions.

This cover illustration was created by artist, Neal Adams, in his graphics studio. The figure makes use of prominent maps of the USGS and the maps of Marie Tharpe. Seafloor spreading isochrons and fracture zones are like multiple railroad tracks that lead backward in time to their points of origin. Continental plates were moved back along those very lines, era by era, across ocean depths, until, 180 million years ago, all the oceans disappear, and only the continental plates remain.

Perhaps, the theory of a growing Earth is an idea whose time has finally come.

Editors

Malte C. Ebach is a systematic biogeographer at the *Laboratoire Informatique et Systématique* (LIS), Université Pierre et Marie Curie (UMR 5143), Paris. Malte's work is primarily on the role of systematic and biogeographical classification in comparative biology that includes several palaeobiogeographical studies involving trilobites, the taxonomic group of his choice. In 2006, Malte helped establish the Systematic and Evolutionary Biogeographical Association (SEBA) and was its firs Chair. Besides the odd study on trilobites, he is deeply interested in the scientifi endeavours of Johann Wolfgang von Goethe and 'Goethe's Way of Science.'

Ray Tangney is head of cryptogamic botany and curator of bryophytes at Amgueddfa Cymru–National Museum Wales in Cardiff. His primary interests are the systematics, evolution, and biogeography of the pleurocarpous mosses. His research is mainly specimen-based taxonomic study, augmented by DNA sequence data, addressing problems of classification and evolution. He also undertakes empirical and theoretical studies on the biogeography of mosses and has a strong interest in spatial patterns in the distributions of related taxa and on branching in mosses, applying techniques of architectural analysis to problems of branching pattern. He has extensive fiel experience in New Zealand and its sub-Antarctic regions, Australia and New Caledonia, and is the author of 20 research papers. He is undertaking continuing work on a monograph of the Lembophyllaceae and he is contributing treatments of the Polytrichaceae and Lembophyllaceae to the Moss Flora of New Zealand project. He is currently co-editing a book on the systematics and evolution of pleurocarpous mosses. Dr. Tangney received his Ph.D. from the University of Otago in 1994 and, after teaching there for eight years, moved to Wales in 2000. He has served on the Council of the Systematics Association and the British Bryological Society, and he is a member of the International Association of Bryologists.

Contributors

Neal Adams
Continuity
New York, New York

John R. Grehan
Buffalo Museum of Science
Buffalo, New York

David J. Hafner
New Mexico Museum of
  Natural History
Albuquerque, New Mexico

Bernhard Hausdorf
Zoologisches Museum der
  Universität Hamburg
Hamburg, Germany

Christian Hennig
Department of Statistical
  Science
University College London
London, United Kingdom

Dennis McCarthy
Buffalo Museum of Science
Buffalo, New York

Lynne R. Parenti
Division of Fishes, Department of
  Vertebrate Zoology
Smithsonian Institution, National
  Museum of Natural History
Washington, D.C.

Brett R. Riddle
Department of Biological Sciences
University of Nevada Las Vegas
Las Vegas, Nevada

Isabel Sanmartín
Department of Systematic Zoology,
  Evolutionary Biology Centre
Uppsala University, Uppsala, Sweden

Tod F. Stuessy
Department of Systematic and
  Evolutionary Botany, Institute of
  Botany
University of Vienna
Vienna, Austria

David M. Williams
Botany Department
The Natural History Museum
London, United Kingdom

Table of Contents

Introduction

Malte C. Ebach

BIOGEOGRAPHY IN A CHANGING WORLD

Comparative biologists have either formed models that explain an organism's history in terms of phylogenetic lineages or genealogies, or have compared their observations in the form of classifications. These practices of comparison and explanation are inherent in systematic studies and have influenced the development and creation of the field of study known as *biogeography*.

There is clearly an unusual breadth of purpose in biogeography today, such that indications for the future direction of studies in this field are confused, to say the least. Will tomorrow's biogeography trace the evolutionary histories of particular organisms case by case, or focus at the level of taxa or even individual gene lineages? Should biogeography be concerned with the examination of ecological regions or biota, or study the classification of endemic areas? Will biogeography be directed at conservation or biodiversity studies, or will it perhaps involve study of human societies and their interactions with nature?

There are a great many interpretations and definitions of biogeography, each with their own slant toward a particular field of study. The molecular biologist may define biogeography to be the study of gene lineages; the ecologist may insist it is about ecosystems and their geographical range; a systematist may favour a biogeography that classifies regions and biota. Biogeography is suffering from something of an identify crisis as the term has become a 'catch-all' to describe the geographical aspect of a particular life-science field rather than existing as an independent fiel of study. It is not surprising, then, that the newly revived field of geographical phylogenetics, called *phylogeography*, which incorporates gene trees, has attracted so many followers so quickly (see Riddle, Chapter 7).

Phylogeography has appeared at a time when many are questioning the relevance of biogeography. This questioning stance may yet prove fruitful if it can be directed toward achieving a greater definition of the aims of biogeography, its historical foundations, its purpose and relevance. Perhaps unwittingly, phylogeography has challenged our perceived ideas about biogeography in several different ways. Phylogeography has emphasised the view that biogeography can have both an ecological and historical aspect at the same time, highlighting the proposed, yet unnecessary division between ecological and historical biogeography (Nelson, 1978). The aims and implementation of phylogeography include non-Croizatian vicariance and dispersal models, bringing into relief the issue of the relevance or otherwise of a division between vicariance and dispersal explanations. Biogeography may make claims to be many things, but at the same time those claims have led to rather confusing internal struggles and the appearance of a contradictory, hence useless, science.

Biogeography sometimes seems caught in a politicised battle between those who call themselves ecologists and those who call themselves systematics, between supporters of vicariance models and dispersalist scenarios, between molecular systematists and morphological systematists, between those who study endemism and those who classify regions, between those who are cladists and those who are pheneticists. Biogeography is a field comprising people from many different backgrounds with different aims and different ideas regarding methods and their application. To understand why we do biogeography and to glean something of its likely future, we need to understand and appreciate its past and foundations (see Williams, Chapter 1, and Parenti, Chapter 2).

Biogeography as a term is relatively new, being coined, but not defined, by Clinton Hart Merriam (1855–1942) (see Merriam, 1892). We can trace "biogeographical" ideas as far back as Aristotle's *Historia Animalia,* the *Historié Naturelle Générale* by Georges-Louis Leclerc, Comte de Buffon (1707–1788) (Buffon, 1761), to the *Flore française* of Jean Baptiste Pierre Antoine de Monet, Chevalier de Lamarck (1744–1829) and Augustin-Pyramus de Candolle (1779–1841) (Lamarck & Candolle, 1805), to the *Essai sur la Géographie* by Alexander von Humboldt (1769–1859) (Humboldt, 1805), *Researches into the Physical History of Mankind* by James Cowles Prichard (1786–1848) (Prichard, 1836), to Friedrich Ratzel's (1844–1904) posthumous work *Allgemeinen Biogeographie* (Müller, 1986). In each and every case, biogeography or zoogeography or phytogeography or general biogeography, is characterized differently by its authors. The systematists and taxonomists practicing biogeography preferred to describe endemic areas in a general classification. The more evolutionary-minded biogeographers, such as Buffon and Ernst Heinrich Philipp August Haeckel (1834–1919), preferred an explicit history of distribution and evolution. It has been the latter tendency that has made the greatest impact on twentieth century biogeography, evidenced in the works of ecologists, palaeontologists, evolutionary biologists, and, most recently, molecular systematists. Biogeography has followed a similar trend in its approaches, as has biological systematics. Systematic methods, theories, and implementations, such as homology, synapomorphies, parsimony, and cladograms, have had a major impact on a *systematic biogeography,* which is the approach that attempts to classify. The same evidence of the importance of the methods theories and implementations applies to evolutionary biology. The field that had endorsed founder dispersal and molecular clocks has founded *evolutionary biogeography* — a field that attempts to explain individual taxic histories by retracing their geographical and evolutionary pathways.

Systematic biogeography vs. *evolutionary biogeography* is a natural tension that exists based on the intentions of its practitioners (Ebach & Morrone, 2005). In fact, the 'two biogeographies' are equally *ecological, historical,* and *paleontological* (see Riddle and Hafner, Chapter 7). Differences between these fields reflect the schisms present in evolutionary biology and systematics. The former aims to 'retrodict' mechanisms within phylogenies (such as in molecular systematics and palaeontology), and the latter aims to provide classifications of endemic areas and biota (as in molecular systematics and palaeontology). In fact, a systematic and evolutionary approach exists in all biological, geographical, and geological, fields. In fields where ancestor-descendant lineages are proposed, so too are sister groups. Where we propose geomorphological mechanisms, we also describe sedimentary structures. Where we predict sympatric or parapatric speciation between populations, we also describe population structure. Where we determine tectonic mechanisms, we also describe mineral morphology and the structure of rocks (petrology).

The divisions within biogeography reflect the same divisions that exist between all historical, that is, non-experimental, science. What makes the study of biogeography unique is the consideration of space, namely in relation to geographical distribution and diversity. As biogeographers we may either explain it or classify it, but we cannot do both at the same time.

Undoubtedly, the majority of people who call themselves biogeographers practise evolutionary biogeography — as evidenced by the explanations, which appear to justify an evolutionary 'purpose' (dispersal mechanisms, migration patterns, identification of species-rich centers of origin or diversification). Conversely, those who pursue classifications include all possible explanations, but also venture to choose among them. The continuing conflict in biogeography is not one of personality, methodology, or implementation — it is of opposing intent. One intention is to show that with explicit histories in play we do not need universal classifications and, conversely, there is the intention to uncover universal classifications that relieve the need for explicit case histories. If we intend to trust our models and decide upon a way to implement a method to generate all possible historical scenarios and choose among them, then we have opted for a *unifying* method or approach. We may, however, decide to classify and identify all possible biota and form a classificatio that may discover *universal* patterns. Under a *universal* approach we acknowledge that some histories may never be discovered or resolved. The choice in approach is entirely up to the biogeographer — either generate models and make assertions based on what these models predict, or discover patterns and then discuss what such fragments of our Earth's history mean.

In re-reading the historical 'biogeographical' literature we find a distinct separation between *systematic* and *evolutionary* biogeography. In order to predict what this may hold for the future of biogeography, we need only look back at the field that have established systematic and evolutionary biogeography. Biological systematics is moving toward larger-scale projects that attempt to diagnose monophyletic groups and their phylogenetic relationships and challenge the acceptance of paraphyletic groups. The modern synthesis, still the theoretical cornerstone of evolutionary biology, is delving deeper into phenetic implementations to measure genetic and non-molecular diversity and distributions. By creating more and more models, which propose mechanisms, evolutionary biology distances itself from classification homology, and the search for monophyletic groups. The same rift is occurring in biogeography between those who wish to understand what areas are and how they may be compared, to those that insist upon modelling of distributional pathways and evolutionary mechanisms. Certain aspects of biogeography, like those of systematics, palaeontology, ecology, and geography, are becoming less and less comparative. Without something to compare, comparative biology becomes meaningless, and systematics and biogeography lose their context in a changing world.

REFERENCES

Buffon, G.L.L. Comte de (1761). *Histoire naturelle générale.* Imprimerie Royale, Paris.

Ebach, M.C. & Morrone, J.J. (2005). Forum on historical biogeography: what is cladistic biogeography? *Journal of Biogeography,* 32, 2179–2183.

Humboldt, A. von (1805). *Essai sur la géographie des PlanLlorente Bousquets, J.tes; Accompagné d'un Tableau Physique des Régions Equinoxiales.* Levrault, Paris.

Lamarck, J.B.P.A. de M. de & Candolle, A.P. de (1805). *Flore française, ou descriptions succinctes de toutes les plantes qui croissent naturellement en France, disposées selon une nouvelle méthode d'analyse, et précédées par un exposé des principes élémentaires de la botanique* (3rd ed.). Desray, Paris.

Merriam, C.H. (1892). The geographical distribution of life in North America with special reference to the Mammalia. *Proceedings of the Biological Society of Washington,* 7, 1–64.

Müller, G.H. (1986) Das koncept der "Allgemeinen Biogeographie" von Friedrich Ratzel (1844–1904): Eine übersicht. *Geographische Zeitschrift,* 74, 3–14.

Nelson, G. (1978). From Candolle to Croizat: comments on the history of biogeography. *Journal of the History of Biology,* 11, 296–305.

Prichard, J.C. (1836). *Researches into the physical history of mankind.* London.

1 Ernst Haeckel and Louis Agassiz: Trees That Bite and Their Geographical Dimension

David M. Williams

ABSTRACT

Ernst Haeckel and Louis Agassiz may seem strange bedfellows; rarely did they write or say anything complementary about each other — their published (and private) comments notable for their hostility. Nonetheless, Haeckel, the proud German 'Darwinian', borrowed a great deal from Agassiz, the archetypical creationist foe, particularly his ideas on the possibilities and potentials of the threefold parallelism: the union of palaeontology, comparative biology (systematics), and ontogeny. Indeed, Haeckel made the threefold parallelism central to his work on 'reconstructing' genealogies — while the ever-frustrated Agassiz (privately) complained that evidence from the threefold parallelism was indeed "mein Resultat!" and supported his kind of 'genealogies'. Agassiz (publicly) discussed his views on Haeckel, especially 'genealogical' trees, and the errors within, his critique centering primarily on palaeontology, the extent and usefulness of the fossil record. But lurking in the background of both men's work was the disturbing evidence of geographical distribution — disturbing, as both saw quite clearly that it, and it alone, could potentially make direct statements concerning the origin of organisms (particularly humans). Haeckel's understanding of geographical distribution was essentially that of the route map, a science he proposed calling *chorology*. Agassiz's understanding of geographical distribution was briefly outlined in his 1859 *Essay on Classification*. He suggested, in passing, that it might indeed constitute a fourth parallelism, alongside palaeontology, systematics, and ontogeny. Agassiz's essay was not much read then and, one suspects, rarely read today. Haeckel's views on geographical distribution were much more widely disseminated, primarily in one of his popular works, *The History of Creation,* the English translation of *Natürliche Schöpfungsgesichte.* As a consequence, much of biogeography today is *chorology.* Given these twists and turns, this paper discusses the views of both Agassiz and Haeckel: their views on 'genealogies', geographic distribution and their union in evolutionary studies.

1

INTRODUCTION

In 1992 Robert Richards published a short but controversial book entitled *The Meaning of Evolution* (Richards, 1992). His focus was a different reading of Darwin, one in which the influence of recapitulation was more profound, influence that hinged on the biology of the 1820s and 1830s, complete with the traditional embryological meaning of 'evolution', rather than the modern more familiar version of species descent with modification. That evolution meant other things in olden times was not a particularly new notion (Bowler, 1976a), but as Richards wrote:

> Historians have understood these two usages of 'evolution' to be quite separate in meaning, like the 'bark' of the dog and the 'bark' of the tree; and to link them would have been comparable to supposing that a tree might bite you (Richards, 1992, p. xiii).

And as Richards pursued his researches, he "became more wary of trees" (Richards, 1992, p. xiii). He was probably right to be wary of trees — once Ernst Haeckel prepared the first explicit genealogies of organisms (Haeckel, 1866), many trees said to depict genealogical relationships made their way into print, their meaning seemingly plain enough but in actuality often quite obscure (Brace, 1981; Nelson & Platnick, 1981; Baez et al., 1985; Craw, 1992; Stevens, 1994; Bouquet, 1995). Today it is a given that trees of organisms really do represent genealogies, pedigrees depicting paths of 'descent with modification'. And it is a given that those same trees might be laid upon a map so 'the wanderings of plants and animals from their first home' appear as route maps and might easily be explained as a result (Hehn, 1885; Gadow, 1913). The map and the genealogy became linked, again by Ernst Haeckel, who prepared a series of plates depicting the relations among humans and their wanderings (of which more later). These 'givens', *actual* genealogies and *actual* route maps have become questionable over time, hence my borrowing from Richards to explore some of these 'trees that bite and their geographical dimension'.

I focus my discussion on two men, Ernst Haeckel (1834–1919) and Louis Agassiz (1807–1873), both significant figures in the history of comparative biology. One might think the two strange bedfellows, Haeckel, champion of materialistic Darwinian evolutionary thinking (Uschmann, 1959, 1983; Hertler & Weingarten, 2001; Richards, 2004, 2005, 2006), Agassiz, a pious, stubborn 'creationist' (Lurie, 1960; Winsor, 1991). Both men commanded attention, respect, and influence during their lives. Both were extraordinarily effective and inspiring lecturers (Dobbs, 2005, for Agassiz; Nyhart, 2002, p. 2 for Haeckel), each leaving a legacy of loyal students (Winsor, 1991, for Agassiz; Franz, 1943–1944, for Haeckel). Both used the popular press to their advantage, promoting their views. Both were accused (correctly in each case) of plagiarism and wrongdoing (Winsor, 1991 for Agassiz; many accounts of Haeckel's embryological plagiarisms exist, from many points of view, two useful modern accounts are Rupp-Eisenreich, 1996, and Richardson & Keuck, 2002). Both were more than delighted to mount explosive and vitriolic attacks on enemies, Haeckel appearing to gain more pleasure from them than Agassiz. And both ended their days in sadness, Agassiz, as a 'living fossil' (Hull, 1973, p. 449; Patterson 1981a, p. 221), struggling but failing to keep his vision of the natural world alive,

Haeckel, the 'tragic Lear-figure', eventually excommunicated from his own museum by Ludwig Plate, his own choice as replacement (Heilborn, 1920; Goldschmidt, 1956; Weindling, 1989, p. 321). Yet the two men are inextricably bound together, somewhat ironically through their uses and abuses of 'evolution', or at least what might be understood or included within that term, and the trouble they both landed themselves in with their views on humans, their relationships and their origins.

PEOPLE THAT BITE: PLAGIARISM
AND THE THREEFOLD PARALLELISM

Haeckel was more than aware of Agassiz's powerful intellect as well as influence "Ich kann es hier nicht unterdrüken, das, Urteil von Ernst Haeckel, seines gefährlichsten Gegners, anzuführen. Ich hatte 1874 Gelegenheit, mit diesem hervorragenden Jenenser Zoologen über sein Verhältnis zu Louis Agassiz zu reden und er gestand mir recht offen, dass er in der Zoologie nur noch einen ebenbürtigen Gegner habe und das sei Agassiz" (Keller, 1929, p. 43)*. Shortly after Agassiz's death, in late 1873, Haeckel excelled himself in *Ziele und Wege der Heutigen Entwickelungsgeschichte* (Haeckel, 1876b). He began slowly with a few pointed criticisms of Agassiz, taking his lead from a critique begun in the contemporaneous and more popular *Anthropogenie,* where he attacks Agassiz's *Essay on Classification* (Agassiz, 1857, 1859; Haeckel, 1874b, p. 116). Then Haeckel picks up steam. First, he accuses Agassiz of plagiarism, for claiming credit for work undertaken by other scientists (Haeckel included, see Ghiselin, 1992; for a detailed account of Agassiz's approach to his students and colleagues as well as his covert plagiarism see Chapter 2 of Winsor, 1991). Haeckel continues by renewing his attack on Agassiz's *Essay on Classification,* following with an attack on the posthumously *Evolution and Permanence of Type* (Agassiz, 1874). He closes by stating quite clearly that Agassiz was nothing but a charlatan and 'racketeer':

> The consequence of this charlatanry is not to be underestimated. But I, at least, see clearly the hoof of Mephisto peeking out from under the black robes of the priest in which the sly Agassiz, with his theatrical decorum and talent for ornament, knew to wrap himself (Haeckel, 1874b, p. 87; translation in Dobbs, 2005, p. 173).

Haeckel's tirade, steeped in a viciousness "happily unusual in natural history" (Marcou, 1896, II, p. 126), was recently summed up as "not simply an insulting counter-argument ... [but] a calculated effort at character assassination" (Haeckel, 1874b, p. 78–86; Dobbs 2005, p. 172). The episode brings to mind Cuvier's *Éloge* on Lamarck (Cuvier, 1832 [reprinted 1984]; Corsi, 1988) and Huxley's summary of Richard Owen's career (Huxley, 1894); the effect of these latter two contributions was to shift focus and identify an enemy. Haeckel need not have bothered, as Agassiz's star was on the wane.

---

* "I cannot suppress Ernst Haeckel's judgement, as his most dangerous opponent. In 1874, I had the opportunity to talk with this outstanding Jena zoologist about his relationship with Louis Agassiz and he confessed to me quite openly that in Zoology he had only one equal opponent and that was Agassiz" (my translation).

Even Haeckel's friends were shocked and not a little angry at the tone of *Ziele und Wege,* Carl Gegenbaur chastising him for the unnecessary venom (Di Gregorio, 1995, p. 270–271). More recently, Michael Ghiselin commented on the event in his review of Polly Winsor's book *Reading the Shape of Nature* (Winsor, 1991), in which she briefly describes Haeckel's attack on the dead Agassiz (Winsor, 1991, p. 54). Winsor suggests that Haeckel may have been "annoyed at the encomiums heaped upon Agassiz after his death" (Winsor, 1991, p. 54; cf. Gladfelter, 2002). Yet, Haeckel's attack may well have been smouldering since the publication of the French translation of Agassiz's *Essay on Classification* in 1869 (Agassiz, 1869a; see Morris, 1997; Lynch, 2003). In this translated edition, Agassiz added three new sections, a discussion of sexual dimorphism, of 'primitive' humans and a critique of Darwinism, the latter being a direct attack on Haeckel, criticising his poor knowledge of the fossil record, particularly its use in support of the genealogies published in *Natü liche Schp̈ fungsgeschichte* (Haeckel, 1868; see also Gould, 1979, 2003 for two accounts of Agassiz's marginalia in *Natü liche Schp̈fungs geschichte* — his marginal comments were evidently inspiration for the additions in the French *Essay on Classification;* Haeckel would not, of course, have seen or known of these words. See below). Some passages and new additions to the French *Essay on Classification* were published separately in the *Revue des cours scientifiques de la France et de l'étranger* (Agassiz, 1869b) — but not the explicit critique of Darwinism. Haeckel made comments on Agassiz's French additions to the *Essay on Classification* in the 7th edition of *Natürliche Schp̈fungs geschichte* (Haeckel, 1879a), but eventually had *Ziele Und Wege Der Heutigen Entwickelungsgeschichte* translated into French, presumably with a view to combating any effects this new translation might have on a French audience. Nonetheless, one would have to search hard to find an attack more savage than Haeckel's — especially as Agassiz was only recently deceased.

Jules Marcou (1896 [1972], II, p. 126, footnote 1), Agassiz's friend and firs biographer, also wrote of Haeckel's attack on the dead Agassiz, naturally dealing with the incident in an altogether more graceful fashion. Marcou drew attention to another appropriation, one in which Marcou (and others, see below) firmly identifie Haeckel's guilt — Marcou had in mind Haeckel's use of Agassiz's threefold parallelism: "It is Agassiz's law, not Haeckel's" (Hyatt, 1894, p. 390).

THE THREEFOLD PARALLELISM: ITS BEGINNING (TIEDEMANN, 1808)?

Tracing the origins of the threefold parallelism — and the 'doctrine of recapitulation', the apparent relation between the results of comparative anatomy (systematics, the 'natural system', classification) and the ontogenetic development of individual organisms — is a formidable task (Kohlbrugge, 1911; Russell, 1916; Shumway, 1932; Meyer, 1935, 1936; Lebedkin, 1936, 1937; Wilson, 1941; Holmes, 1944; Oppenheimer, 1959; Gould, 1977; Mayr, 1994; Müller, 1998). Notwithstanding Kohlbrugge's (1911, p. 448) valiant attempt — he lists 72 possible candidates, beginning in 1797 with contributions from Goethe and Autenrieth — and Meyer's (1935) tentative vision of its origin with Aristotle, one might easily be forgiven for attributing the notion to Agassiz, as Agassiz himself was wont to do (Agassiz, 1844a; Agassiz & Gould, 1848; see Marcou 1896, I, p. 230). While Agassiz acknowledged Tiedemann as the 'true' creator of recapitulation ([Agassiz] 1860–1862, p. 245; see Lurie 1960, p. 286; Tiedemann had taught Agassiz, see Marcou 1896, I, p. 16, Lurie 1960, p. 21) as he had noted a relationship between comparative anatomy, development, and the fossil record (Tiedemann, 1808, p. 73; Russell, 1916, p. 255 footnote 3), Agassiz clearly believed he had discovered something, writing "Das ist mein Resultat!" in the margin of his copy of *Natürliche Schöpfungsgeschichte* (1868, p. 280) against Haeckel's treatment of recapitulation (Gould, 1973, p. 322, 1979, p. 280). In fact, Agassiz's concerns were with the inclusion of 'geological succession' among other forms of evidence:

> But I may at least be permitted to speak of my own efforts, and to sum up in the fewest words the result of my life's work. I have devoted my whole life to the study of Nature, and yet a single sentence may express all that I have done. I have shown that there is a correspondence between the succession of Fishes in geological times and the different stages of their growth in the egg, — this is all (Agassiz, 1862, p. 52).

Agassiz first wrote of 'recapitulation' and its relationship to geology in his *Poisson fossiles,* (Part I, 1844a, p. 81, 102), the relevant passages coming from the first volume. Years earlier, Charles Lyell, in the second volume of his *Principles of Geology,* was able to write:

> There is yet another department of anatomical discovery, to which we must not omit some allusion, because it has appeared to some persons to afford a distant analogy, at least, to that progressive development by which some of the inferior species may have been gradually perfected into those of more complex organization. Tieddemann [sic] found, and his discoveries have been most fully confirmed and elucidated by M. Serres, that the brain of the foetus, in the highest class of vertebrated animals, assumes, in succession, the various forms which belong to fishes, reptiles and birds, before it acquires those additions and modifications which are peculiar to the mammiferous tribe. So that in the passage from the embryo to the perfect mammifer, there is a typical representation, as it were, of all those transformations which the primitive species are supposed to have undergone, during a long series of generations, between the present period and the remotest geological era (Lyell 1832, p. 62–63 [1991 reprint*]).

It may well have been Agassiz's students, friends, and teachers**, who were responsible for giving more substance to Agassiz's ideas (see also Russell, 1916, p. 255). Alpheus Hyatt (1894, p. 390) and Jules Marcou (1896, II, p. 126), both students of Agassiz, drew attention to Haeckel's appropriation (Russell 1916, p. 255; Gould, 1979, p. 280; Janvier 1996 [1998], p. 313; Williams & Ebach, 2004, p. 691), and Bronn, one of Agassiz's teachers, gave him credit for recognising this particular threefold parallelism, with the inclusion of fossils (Bronn 1858, p. 103, 1861, p. 534). The significance of Agassiz's work was rendered more profound by Joseph

---

* The texts in various versions differ.
** See Winsor (1991, p. 35) for a list of Agassiz's students.

Le Conte, another Agassiz student and friend (Le Conte, 1903). Writing some 10 years after Agassiz's death:

> I know that many think with Haeckel that biology was kept back half a century by the baleful influence of Agassiz and Cuvier; but I can not think so. The hypothesis [the evolution hypothesis] was contrary to the facts of science, *as then known and understood.* It was conceived in the spirit of baseless speculation, rather than of cautious induction; of skilful elaboration, rather than of earnest truth-seeking. Its general acceptance would have debauched the true spirit of science ... . The ground must first be cleared ... and an insuperable obstacle to hearty rational acceptance must first be removed, and an inductive basis laid (Le Conte, 1888, p. 33, 1905, p. 33).

Naturally, Le Conte considered Agassiz the person who cleared the ground, as

> ... the only solid foundation, of a true theory of evolution ... is found in ... the method of comparison of the phylogenic and the embryonic succession, ... and the laws of embryonic development (ontogeny) are also the laws of geologic succession (Le Conte, 1888, p. 33, 1905, p. 33).

Le Conte held the view that "no one was reasonably entitled to believe in the transformation of species prior to the publication of the work of Agassiz" (Lovejoy, 1909a, p. 501; Lurie, 1960; see Le Conte, 1903, p. 151). Be that as it may, Haeckel, early on as noted above, embraced the notion of 'recapitulation' — or the effica y of a threefold parallelism providing evidence. As early as 1863 Haeckel wrote for the first time his perception of the threefold parallelism and how together they present the strongest evidence possible for evolution:

> [...] the threefold parallelism between the embryological, systematic, and palaeontological development of organisms, this threefold step-ladder, I think is one of the strongest proofs of the truth of the theory of evolution (Haeckel, 1863, p. 29; translation modified from Hoßfeld & Olsson, 2003, p. 296; also see Heberer, 1968, p. 58 and endnote ii).

Later Haeckel was more explicit:

> The laws of inheritance and adaptation known to us are completely sufficient to explain this exceedingly important and interesting phenomenon, which may be briefly designated as the *parallelism of individual, of paleontological and of systematic development* (Haeckel, 1876a, I, p. 313, emphasis in the original, see 1872, I, p. 471, for Haeckel's first note).

Whatever relation might emerge between individual, paleontological and systematic development (see below), Haeckel did not see a role for biogeography *as primary evidence,* in spite of the fact that Darwin and Wallace required a geographical dimension for the mechanism of species descent they proposed (Richardson, 1981). To the contrary, Agassiz's interest in the possibilities of geographical differentiation and its meaning extended his entire career.

ERNST HAECKEL AND DARWINISM

'E pur si muove'

Haeckel was lecturing on Darwin as early as 1863, giving his first public lecture at the age of 29 to the 38th meeting of the *Society of German Naturalists and Physicians* in Stettin (Haeckel, 1863). Even the presentation's title, *Über die Entwicklungstheorie Darwin's**, would be enough to cause concern in later years. One translation, 'On Darwin's theory of development', (Nyhart, 1995, p. 129) suggests that evolution is a form of development, while an alternative translation, 'On Darwin's theory of evolution' (Hoßfeld & Olsson, 2003, p. 296), avoids the interpretative issues altogether. But as Nyhart (1995, p. 129) points out, Haeckel could easily have used the more appropriate *'Descendenztheorie'* for 'evolution' — a word he did use later (Haeckel, 1868). Nevertheless, Haeckel began as he intended to go on, dividing the world neatly into righteous Darwinists ("Development and progress!" — "Entwickelung und Fortschritt") and intransigent conservatives ("Creation and species!" — "Schöpfung und Species!") (Haeckel, 1863, p. 18**).

Slowly recovering from the sudden death of his wife in early 1864 (Weindling, 1989, p. 319; Desmond, 1994, p. 349; Richards, 2004, 2005, 2006), Haeckel began assembling what eventually became *Generelle Morphologie,* an enormous book containing all his ideas, a book that would spew "fire and ash over the enemies of progress, and radically alter the intellectual terrain in German biological science" (Richards, 2006). *Generelle Morphologie* had its origin in lecture notes taken by his students, the book being written in 1865–1866, finally being published toward the end of 1866 (Ulrich, 1967, 1968; Uschmann, 1967a). Yet *Generelle Morphologie* was not well received, and Haeckel almost immediately embarked upon another project to render his ideas more accessible, in a more popular format. Of *Generelle Morphologie* and its failure to impress, Radl wrote "The German professors treated the book as a belated offshoot of the long discarded *Naturphilosophie,* and paid little attention to it" (Radl, 1930, p. 122–123). Regardless of its effect, one remains impressed with Haeckel's sense of commitment in promoting Darwin's cause: two thirds of the way down the title page of both volumes, ranged to the right in small capitals, were printed the words: "E pur si muove".

Haeckel was again lecturing on Darwin in 1867–1868; these forming the basis of *Natü liche Schöpfungsgeschichte* (1868), his first popular book, which was to summarise the complexities of his *Generelle Morphologie.* It was *Natü liche Schp̈fungsgeschichte* that had the greater impact on German science as well as the general public — if not a huge part of the reading public via its various translations (Nordenskiöld, 1936, p. 515). The book's influence can be appreciated by the 12

---

* Richards (2005, p. 1) provides a newspaper account of the event, where the reporter writes of the "huge applause [that] followed this exciting lecture". The lecture was later reprinted, with minor amendments, in a collection of articles (Haeckel, 1902); the amended version was reprinted in Heberer (1968).

** Haeckel later wrote more inflammatory expressions of his 'battle': "On the one side spiritual freedom and truth, reason and culture, evolution and progress stand under the bright banner of science; on the other side, under the black flag of hierarchy, stand spiritual slavery and falsehood, irrationality and barbarism, superstition and retrogression" (Haeckel, 1874a, pp. xiii-xiv).

German editions (1868–1920)* and the numerous translations — at least 25 (Oppen-heimer, 1987) — many in English**, editions that Haeckel continually updated throughout his life.

As a working scientist, Haeckel published many technical reports and mono-graphs throughout his life (Krumbach, 1919, reprinted in Heberer, 1968, p. 15–22; Tort, 1996, p. 2114–2121) but managed to find time to produce other popular books including *Anthropogenie* (1874b), dealing with human evolution (of which more later). Some 28 years after *Generelle Morphologie,* Haeckel published *Systematische Phylogenie,* a revision of the systematics and genealogies of the entire living world — a book that went to three volumes and 1800 pages, taking over three years to publish (Haeckel, 1894–1896). One might see these volumes as an attempt to update and possibly eclipse *Generelle Morphologie.*

TREES THAT BITE: HAECKEL'S GENEALOGICAL OAKS AND STICK 'TREES'

While the booming of guns at the Battle of Königgrätz in 1866 announced the demise of the old Federal German Diet and the beginning of a new splendid period in the history of the German Reich, here in Jena the history of the phylum [*Stammesge-schichte*] was born (Volkmann, 1943, p. 85, translation from Gasman, 1971 [2004], p. 18).

... whatever hesitation may not unfrequently be felt by less daring minds, in following Haeckel in many of his speculations, his attempt to systematise the doctrine of Evo-lution and to exhibit its influence as the central thought of modern biology, cannot fail to have a far-reaching influence on the progress of science (Huxley, 1878, p. 744).

While none of Haeckel's genealogical trees appear in *The Hierarchy of Life* (Fernholm et al., 1989), a book whose aim was to "... summarise[s] the progress we have made towards a tree of life ... a goal hardly attempted since Haeckel" (Patterson, 1989, p. 486), one of his illustrations does grace the dust jacket of its successor,

---

* Although 13 editions are listed below, the 13th was produced after Haeckel's death.
** Dates of editions as follows: 1st, 1868, 2nd 1870, 3rd 1872 (used for the first English translation), 4th 1873, 5th 1874, 6th 1875, 7th 1879, 8th 1889, 9th 1898, 10th 1902, 11th 1909, 12th 1911, 13th 1923 (Tort, 1996, p. 2115; Nelson & Ladiges, 2001, p. 404); for English translations, 1st 1876, 2nd 1876, 3rd 1883, 4th 1892, (2nd printing 1899), 5th 1906, 6th 1925, and reprinted as late as 1990 by Chadwyck-Healey Ltd, Cambridge. An Italian edition was published in 1892 (*Storia della creazione: conferenze scientifico-populari sulla teoria dell-evoluzione e specialmente su quella di Darwin, Goethe e Lamark.* Trad. D. Rosa. Torino: Unione Tipografica Editrice), and various French editions were published (1874, *Histoire de la création des etres organisés d'après les lois naturelles.* C. Rienwald, Paris, 1st ed; 1877, *Histoire de la création des êtres organisés d'après les lois naturels.* Traduit de l'allemand par le Dr. Ch. Letourneau et revue sur la septième édition allemande...; et précéde d'une introduction biographique par Charles Martin. C. Rienwald, Paris, 2nd ed; 1879, *Histoire de la création des êtres organisés d'après les lois naturels.* Traduit de l'allemand par le Dr. Ch. Letourneau et revue sur la septième édition allemande. Schleicher, Paris, 3rd ed; 1909, *Histoire de la création des êtres organisés d'après les lois naturelles.* Traduit de l'allemand par le Dr. Ch. Letourneau et revue sur la septième édition allemande, Schleicher, Paris). See Roger (1983, p. 165) for some commentary on the various translations.

Assembling the Tree of Life (Cracraft & Donoghue, 2004). Not surprisingly, the editors chose the *Monophyletischer Stammbaum der Organismen,* or the monophyletic tribe of organisms, a reproduction of the first plate in Haeckel's *Generelle Morphologie* (Haeckel, 1866). Another of Haeckel's trees is found in the introductory essay to *Assembling the Tree of Life* (Cracraft & Donoghue, 2004, Figure 1.2, taken from Haeckel, 1866, the inset of Taf. VII), as one of the four believed to have contributed most to the developing views on how best to represent the living world (the three other diagrams are taken from Darwin, 1859; Zimmermann, 1931, p. 1004, Figure 179; Hennig, 1966, p. 91). These four trees are seen to collectively mark the passage of the birth of 'phylogenetics', from Darwin via Haeckel, Zimmermann and finally Hennig to its present 'maturity', in the algorithms of the numerical taxonomists (Cracraft & Donoghue, 2004, pp. 1–3) — a journey of less than 150 years. That depiction may well be a simple caricature (see, for example, Richards, 1992), but Haeckel remains, if not father, then midwife to the concept of phylogenetic trees. In the hands of Haeckel, the union of Darwin, Bronn, and Schleicher (see below) produced many offspring, mostly in the form of trees attempting to represent the genealogical relations among organisms, their phylogenetic history.

HEINRICH GEORG BRONN: TRUNKS AND TWIGS

Natura doceri

In 1850 the Paris *Academie des Science* offered a prize to anyone satisfactorily answering a question posed concerning the fossil record and what it represents of the changing life that inhabits the Earth and has inhabited it in the past. The task was to present an essay

> to study the laws of the distribution of fossil organisms in the different sedimentary strata according to the order of their supposition; to discuss the question of their successive or simultaneous appearance or disappearance; to examine the nature of the relations between the present and the former states of the organic world (*Comptes-Rendus* 30, pp. 257–260, 1850; translation from Rudwick, 1972 [1985], p. 219; Nelson, 1989, p. 64; see also introduction to Bronn, 1859a, p. 81 and Laurent, 1997).

The opportunity offered by the *Academie des Science* reflects a lingering interest in the notion of the transformation of species, inspired, if not initiated, by Lamark in the early 1800s (Laurent, 1987, 2001). The award, a gold medal to the value of 3000 francs, was announced in 1857 (*Comptes-Rendus* 44, pp. 167–169, 1857) and presented to the eminent palaeontologist Heinrich Georg Bronn (1800–1862, portrait in Burkhardt et al., 1993, opposite p. 89, and Seibold & Seibold, 1997, p. 521, Abb. 2); Bronn would be first to translate Darwin's *Origin of Species* into German. The prize winning essay was first published in its original German in 1858 as *Untersuchungen über die Entwickelungs-Gesetze der organischen Welt* (Bronn, 1858), a full French translation not appearing until a few years later, in 1861 (Bronn, 1861). On the title page of each full edition were the words *Natura doceri,* 'Being taught by nature,' a phrase that would have appealed to Agassiz (Winsor, 1991). An English and French translation of the concluding section, 'On the Laws of Evolution of the Organic World during the Formation of the Crust of the Earth', was published a few months prior to Darwin's *Origin of Species* (Bronn, 1859a, b).

Bronn's monograph included a great many summary diagrams derived from the fossil record of various groups of animals, diagrams similar to Agassiz's (Figure 1.1). Bronn noted that such systems could be generalised by a branching diagram, relating various groups of animals to each other, which he included (Figure 1.2a, reproduced from Bronn, 1861, p. 900 after Bronn, 1858, p. 481; see also the dust jacket of Bowler, 1976b and his plate X, Uschmann, 1967b, p. 15, Bowler, 1988 [1992], p. 55 and Craw, 1992, p. 69, Figure 1A, for other reproductions):

> Veut-on représenter cet état de chose par une figure, il fait se figurer le système comme une arbre, où la position plus ou moins élevée des branches correspond à la perfection relative de l'organisation, d'une manière absolue et sans tenir compte de la position plus ou moins élevée des rameaux sur la même branche (Bronn, 1861, p. 899, 1858, p. 481).

The tree has a main trunk, with a number of main branches, A through G, representing invertebrates, fish, reptiles, birds, mammals, and man. A further series of subsidiary branches are labelled with lower case letters — *a* to *m* — representing species at different levels of development and time of appearance in the geological record — this diagram was to have some influence on Haeckel. The design of Bronn's tree is very much like some of Haeckel's, whether the *stammbaum* is Oak-like (Figure 1.2b) or stick-like (Figure 1.2c).

Bronn argued that the fossil remains recorded in various strata show the replacing of earlier groups of organisms with later ones, the later groups better adapted to local environments. Bronn maintained that the fossil remains show conclusively a constant progress from the early simple forms of life to the more complex ones. Yet, he also concluded that the fossil record does not allow direct access to the successive appearance of the various species and, more significantl , does not provide any proof of the transmutation of species (Baron, 1961; Junker, 1991; Laurent, 1997; Seibold & Seibold, 1997). Interestingly, Bronn noted that to distinguish between species fixity or their transformation amount to a commitment of belief, the problem of distinguishing between the two was beyond empirical resolution (Bronn, 1858). Bronn returned to this theme when he added an additional chapter to the German translation of Darwin's *Origin of Species* (Bronn, 1860), the version that was read by Haeckel (Richards, 2005).

Haeckel began reading the German translation of Darwin's *Origin* in the summer of 1860, taking it up again in November 1861 (Richards, 2002). Encouraged by Darwin*, Bronn added his own 15th chapter outlining some of the difficulties he had with Darwin's thesis. Bronn's additional chapter had a significant impact on Haeckel — and German evolutionary thinking from there on (Junker, 1991). For, while Bronn states his enthusiasm for Darwin's general thesis, he suggests that it remains simply a hypothesis, one "possible scenario of life's history" (Richards, 2005):

---

* Although see the comments in Burkhardt et al. 1993, p. 102–103 and 407–409, especially footnote 1.

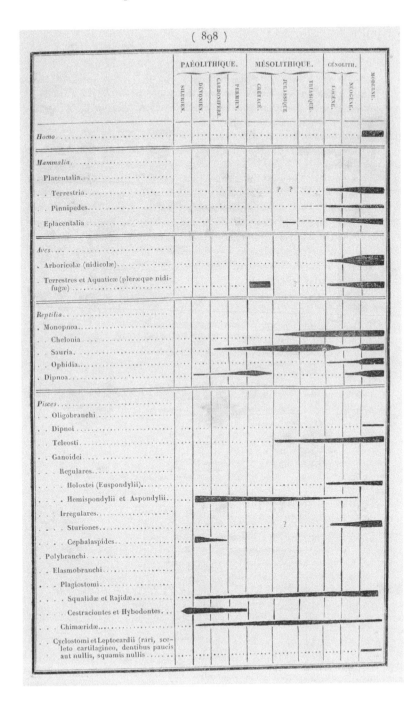

FIGURE 1.1 Summary diagram of the fossil record of various groups of animals (after Bronn 1858, p. 480, 1861, p. 898).

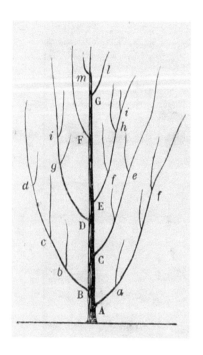

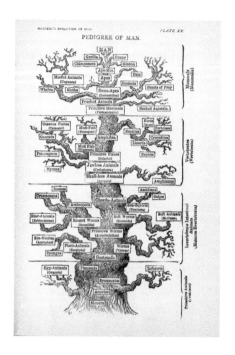

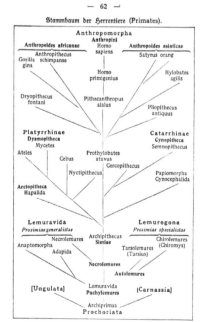

FIGURE 1.2 (a) Branching diagram relating various groups of animals to each other (from Bronn, 1858, p. 481, 1861, p. 900); (b) oak-like *stammbaum* (Haeckel 1879b, plate XV); (c) stick-like genealogy (Haeckel 1907, p. 62).

We have therefore neither a positive demonstration of descent nor — from the fact that [after hundreds of generations] a variety can no longer be connected with its ancestral form (*Stamm-Form*) — do we have a negative demonstration that this species did not arise from that one. What might be the possibility of unlimited change is now and for a long time will remain an undemonstrated, and indeed, an uncontradicted hypothesis (Bronn, 1860, p. 502, 1863, p. 533, translation from Richards, 2006).

Haeckel saw a way of solidifying Darwin's suggestions of the genealogical connection of all organisms by marrying the systematic arrangement of organisms, as revealed by the hierarchical 'natural system', with a graphic representation of genealogy, a pedigree of species. Hence visually, as well as positively, Haeckel could provide both an account of as well as a depiction of which species gave rise to others. Thus for Haeckel, of course, his genealogies *did* represent the transformationist view, graphically and literally. Haeckel also required some independent evidence, a way of tackling Bronn's objections. He found that in other trees, those of the linguist August Schleicher.

SCHLEICHER: LINGUISTICS AND TREES

The linguist August Schleicher (1821–1868) worked at Jena University, becoming a good friend of Ernst Haeckel, who insisted he read Darwin's *Origin,* the German translation made by Bronn (Koerner, 1989; Alter, 1999; Di Gregorio, 2002; Richards, 2002). Schleicher was immensely impressed, responding almost immediately with an open letter to Haeckel, *Die Darwinische Theorie und die Sprachwissenschaft* (Schleicher, 1863, later translated as *Darwinism tested by the Science of Language* in 1869, after Schleicher's death). Schleicher argued that contemporary languages had also undergone a process of change, not too dissimilar from that Darwin suggested for organisms: evolutionary theory *confirmed* language descent, rather than suggesting it. Schleicher had already anticipated such a development, as is evident from his 1860 book and the earlier 1853 article (Schleicher, 1853, 1860):

> These assumptions [the origins of an Indo-European language family], deduced logically from the results of previous research, can best be depicted by the image of a branching tree (Schleicher, 1853, p. 787, translation from Koerner, 1987, p. 112).

In *Die Darwinische Theorie,* Schleicher referred to the tree Darwin provided in the *Origin,* noting that it was a purely hypothetical construct, containing no real species, either at the tips or nodes. This he compared to a tree he had constructed, one depicting the Indo-German languages and included as a figur appended to the 1863 essay (Figure 1.3a, reproduced from Schleicher, 1863, also in Alter, 1999, p. 75, Figure 4.1, and Atkinson & Gray, 2005, p. 518, Figure 2). As early as 1850, Schleicher suggested using a tree-like diagram (a *Stammbaum*) for representing the development of languages, publishing the first in 1853

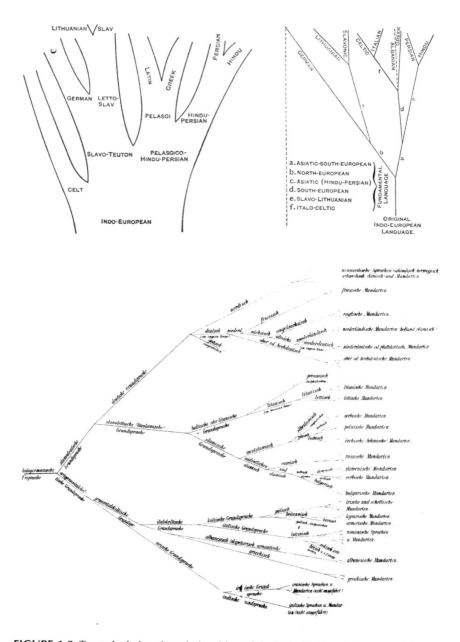

FIGURE 1.3 Trees depicting the relationships of the Indo-German languages: (a) reproduced from Schleicher (1863, also in Alter, 1999, p. 75, Figure 4.1; Atkinson & Gray, 2005, p. 518, Figure 2); (b) a tree-like diagram (a *Stammbaum*) representing the development of languages from Schleicher (1853); and (c) another tree from Schleicher (1860).

(Figure 1.3b, Schleicher, 1853) and another in 1860 (Figure 1.3c, Schleicher, 1860). Schleicher has received credit for introducing the tree as a graphic way of representing genealogies of languages (Richards, 2002; although see Koerner, 1987, and Traub, 1993, for further details). Schleicher's three trees have a modern language named at each tip with the branches and nodes labelled for their common origin (Figure 1.3b and 1.3c). Thus, in both the 1853 and 1860 tree, there are branches for the Persian and Hindu language, for example, linked by a node labelled Hindu–Persian (1853) or Asiatic (Hindu–Persian) (1860). Schleicher was convinced that the evolution of languages provided definit ve evidence for the evolution of man and a way of tracing their development. He referred to the problems of identifying transitional organisms from few fossil remains, noting that there were far more linguistic fossils than there were geological fossils (Richards, 2002), a view that Haeckel later echoed:

The former [philology] can ... adduce far more direct evidence than the latter [geology], because the palaeontological materials of Philology, the ancient monuments of extinct tongues, have been better preserved than the palaeontological materials of Comparative Zoology, the fossil bones of vertebrates ... (Haeckel, 1879a, p. 24*).

In his second contribution to the evolution of man, *Ueber die Bedeutung der Sprache fü die Naturgeschichte des Menschen,* Schleicher criticises the available morphological evidence for relating various humans as superficial, and suggests that language provided a 'higher criterion, an exclusive property of man' (Schleicher, 1865, pp. 18–19, translation from Richards, 2002). Schleicher's arguments for linguistic superiority are remarkably similar to those offered today for DNA sequence data, the universal 'higher criterion'.

As Richards notes:

... Schleicher's greatest and lasting contribution to evolutionary understanding may simply be the use of a *Stammbaum* to illustrate the descent of languages ... . Haeckel quite obviously took his inspiration from his good friend Schleicher. And Haeckel's *Stammbaüme* have become models for the representation of descent ever since (Richards, 2002).

And as Alter states more boldly: "... the historical significance of their friendship is enormous" (Alter, 1999, p. 117). Schleicher died at 48 years old in 1868, two years after Haeckel published his great work (Haeckel, 1866). Haeckel would later include a tree of the "Pedigree of the Indo-Germanic languages" in many of his books, beginning with the 2nd German edition of *Natü liche Schpfungs geschichte* (1870, p. 625) and all subsequent editions, both German and English, of *Anthropogenie* (e.g., Figure 1.4, after *The Evolution of Man* 1883, II, p. 23).

---

* In 1859 and 1863 a two-volume work entitled *Essai de paléontologie linguistique* was published by Adolphe Pictet, a distant relative of the palaeontologist François Jules Pictet (Wells, 1987, pp. 47–48).

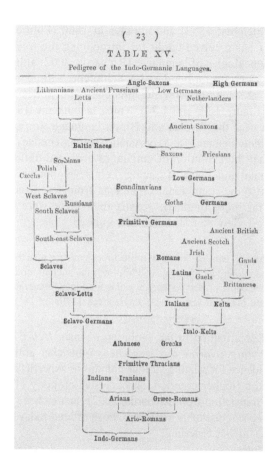

FIGURE 1.4 A "Pedigree of the Indo-Germanic languages" (Haeckel 1883, II, p. 23).

HAECKEL AND PALAEONTOLOGICAL TRUTH

Haeckel was able to evaluate the palaeontological evidence provided by Bronn, as well as utilise the graphic representations given by both Bronn and Schleicher. That is, a union of palaeontology and genealogy would illustrate Darwin's views exactly — or so it seemed.

Haeckel invented the word phylogeny (Haeckel, 1866, I, p. 57, II, p. 301: "Generelle phylogenie oder Allegemeine Entwickelungsgeschichte der organischen Stämme") for what he later described as the "tribal history, or 'palaeontological history of evolution'", adding for precision, "Phylogeny includes palaeontology and genealogy" (Haeckel, 1874a, p. 710, 1883, II, p. 460, see Haeckel, 1866, II, p. 305). With palaeontology firmly in mind as the prime source of evidence for genealogy, Haeckel speculated on the role of individual development:

This palaeontological history of the development of organisms, which we may term *Phylogeny,* stands in the most important and remarkable relation to the other branches of organic history or development, I mean that of individuals, or Ontogeny. On the whole, the one runs parallel to the other. In fact, the history of individual development, or Ontogeny, is a short and quick recapitulation of palaeontological development, or Phylogeny, dependent on the laws of Inheritance and Adaptation (Haeckel, 1876a, p. 10–11).

Regardless of any truths to this assertion, classifications, if understood 'properly', should represent the phylogenetic relationships of organisms — one (genealogy) being derived from the other (classification). Haeckel connected classificatio directly with the "facts of palaeontology", facts that even then were variously interpreted, especially as they related to issues of 'transformation' (see Bronn above). Haeckel had early on made some efforts to convert the 'natural system' of classifi cation into a genealogical scheme, in the first part of his monograph on the *Radiolaria,* in a table entitled *Genealogische Verwandtschaftstabelle der Familien, Subfamilien und Gattungen der Radiolarien* (Figure 1.5; Haeckel, 1862, p. 234*). The 1862 monograph had its origins in Haeckel's earlier *Habilitationschrift* studies (Haeckel, 1860, 1861), where he described many new genera and species, and although at that time he pondered their genealogical relationships, he resisted any mention of how Darwin's ideas might influence classification. Nevertheless, along with the 'genealogical' table, the 1862 monograph included a footnoted discussion on Darwin's *Origin* (Haeckel, 1862, pp. 231–232).

Haeckel's *Generelle Morphologie* was published in two volumes; the second included a *Systematische Einleitung in die allgemeine Entwickelungsgeschichte,* a detailed description and classification of all life, with eight plates depicting various *Stä mbaume* or pedigrees, the first diagrams of their kind (Haeckel, 1866, "mit acht Genealogischen Tafeln"; listed in Table 1.1). *Generelle Morphologie* appeared in just one edition and was never translated**, although a condensed German language edition was eventually published 40 years later (Haeckel, 1906), excluding the eight *Stä mbaume,* and Heberer (1968; see also Uschmann, 1967a and Ulrich, 1967, 1968) published large extracts for a collection of Haeckel's work celebrating the 100th anniversary of *Generelle Morphologie***.

In fi e of Haeckel's plates (Table 1.2, Tafs III, IV, V, VI and VII) there are two trees per plate, a larger, more detailed tree with many more named nodes, and a

---

* Haeckel's monograph was finally completed in 1888, extending to three volumes (Haeckel, 1862, 1887a, b). In between times, he presented a series of genealogies and charts capturing the interrelationships of various Radiolarian groups derived from the Challenger material (Haeckel, 1887a) and presenting a final summary in the form of a chart and diagram in *Systematische Phylogenie* (Haeckel, 1894, pp. 207–208; see Aescht, 1998).

** Not for want of trying. Every effort was made by Huxley to translate the book, finally deciding that it was "too profound and too long" (from Darwin in Desmond, 1982, p. 155). Richards (2006) deals with this episode in detail.

*** Heberer (1968) includes reproductions of only Taf. I–III, V–VIII, noting "Es Folgen auf den Seiten 269–274 den acht genealogosichen Tafeln, die dem Bd. 2 der "Generelle Morphologie" von Haeckel beigegeben wurden, die Nrn. I, II, III, IV, VI, VIII. Massgebliche Fachleute haben dazu bemerkt, man müsste mit Erstaunen feststellen, dass Haeckel schon damals die phylogenetischen Beziehungen der Ogranismen in wesentlichen Zügen richtig erfasst habe." (Heberer, 1968, p. 268). Although Haeckel's genealogical trees have been reproduced on many occasions, Papavero *et al.* (1997) reproduce all eight genealogies (Papavero et al., 1997, pp. 254–261) from the 2nd French edition of *Natü liche Schpfungs - geschichte, Histoire de la Creation....*

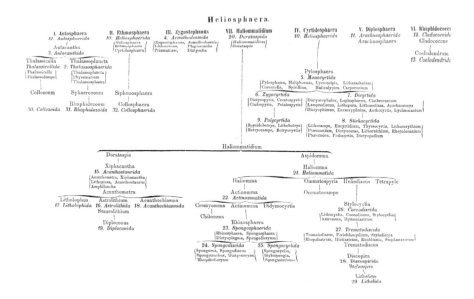

FIGURE 1.5 An early classification of Haeckel's, as a genealogical scheme (Haeckel, 1862, p. 234).

smaller inset, presumably representing the morphological relationships. Two of the fi e (Table 1.2, Taf. IV and Taf. VII) include explicit palaeontological eras ("palaeontologisch begründet") appended to the axis of the main tree, which has been interpreted as adding common descent to the fossil record (Patterson, 1983, p. 8).

Natü liche Schöpfungsgeschichte went through twenve editions, the text and illustrations being modified accordingly. The 1st German edition has eight plates, different from those published in *Generelle Morphologie* (Table 1.1). In all of the later editions (8th–12th), the pedigrees, *Stä mbaume* and genealogies are represented in a simpler format, as charts within the text (e.g., Figure 1.6), the 'branches'

TABLE 1.1

Plates from Generelle Morphologie, Systematische Einleitung in die allgemeine Entwickelungsgeschichte (Haeckel, 1866)

Generelle Morphologie, Systematische Einleitung in die allgemeine Entwickelungsgeschichte (1866)

Taf. I	Monophyletischer Stammbaum der Organismen
Taf. II	Stammbaum des Pflanzenreich
Taf. III	Stammbaum des Coelenteraten oder Acalephen (Zoophyten)
Taf. IV	Stammbaum des Echinodermen palaeontologisch begründet
Taf. V	Stammbaum des Articulaten (Infusorien, Würmer und Arthropoden)
Taf. VI	Stammbaum des Mollusken (Molluscoiden und Otocardien)
Taf. VII	Stammbaum des Wirbelthiere palaeontologisch begründet
Taf. VIII	Stammbaum des Säugethiere mit Inbegriff des Menschen

TABLE 1.2
Plates from the 1st–9th Editions of Haeckel's *Natürliche*
Schöpfungsgeschichte **and the English Translation,** *The History of Creation,*
or the Development of the Earth and Its Inhabitants by the Action of Natural
Causes **(Haeckel, 1876)**

Natürliche Schöpfungsgeschichte (1868) 1st Edition

Taf. I	Einstämmiger oder monophyletischer Stammbaum der Organismen
Taf. II	Einheitlicher oder monophyletischer Stammbaum des Pflanzenreichs palaeontologisch begründet
Taf. III	Einstämmiger oder monophyletischer Stammbaum des Thierreichs
Taf. IV	Historisches Wachsthum der jechs Thierstämme. Siehe die Erflärun
Taf. V	Stammbaum des Oliedfüsser oder Arthropoden
Taf. VI	Einheitlicher oder monophyletischer Stammbaum des Wirbelthierstammes palaeontologisch begründe
Taf. VII	Stammbaum der Säugethiere mit Inbegriff des Menschen
Taf. VIII	Stammbaum der Menschen-Arten oder Classen.

Natürliche Schöpfungsgeschichte (1870, 2nd Edition — 1879, 7th Edition)

Taf. IV	Einheitlicher oder monophyletischer Stammbaum des Pflanzenreichs palaeontologisch begründet
Taf. V	Historisches Wachsthum der sechs Thierstämme
Taf. XII	Einheitlicher oder monophyletischer Stammbaum des Wirbelthierstammes palaeontologisch begründet
Taf. XV	Hypotheische Skizze des monophyletischen Ursprungs und der Verbreitung der 12 Menschen-Species von Lemurien aus über die Erde

Natürliche Schöpfungsgeschichte (1889, 8th Edition — 1898, 9th Edition)

Taf. XX	Hypotheische Skizze des monophyletischen Ursprungs und der Verbreitung der 12 Menschen-Species von Lemurien aus über die Erde

The History of Creation, or the Development of the Earth and Its Inhabitants by the Action of Natural Causes (1876)

Taf. V	Single-stemmed monophyletic pedigree of the vegetable kingdom based on palaeontology
Taf. VI	Historical Growth of the six great stems of animals
Taf. XIV	Single or monophyletic pedigree of the stem of the back-boned animals based on palaeontology
Taf. XV	Hypothetical Sketch of the Monophyletic Origin and the Extension of the 12 Races of Man from Lemuria over the Earth

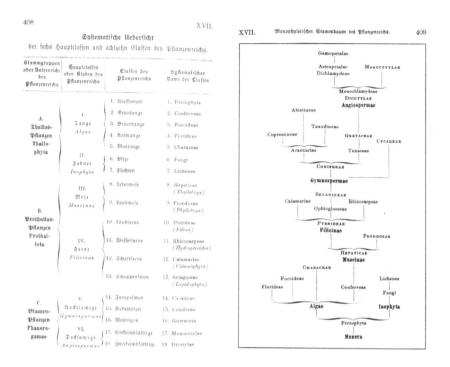

FIGURE 1.6 (a) Illustration of "Systematische Ueberfischt der sechs hauptklassen und acht-zehn klassen des Pflanzenreich" and (b) "Monophyletischer Stammbaum des Pflanzenreich" from Haeckel (1879a, pp. 408–409).

becoming simple lines directly connecting taxon names, the lines relating taxa specified in an accompanying synoptic table (*Systematiche Übersichten*).

The 2nd through to the 7th editions have only four plates (Table 1.2), while the 8th and 9th editions have but a single plate (Table 1.2)*. Three of the plates in the 2nd to 7th edition are reproductions from the 1st edition (compare the lists in Table 1.1). Of the eight plates in the 1st edition, plates II and VI are said to be "palaeontologisch begründet;" these are the retained plates for the 2nd to the 7th edition, both included in the English editions (Table 1.2). The fourth plate, however, is unique and was not published in the 1st edition. It represents Haeckel's tree of humans and their wan-derings, *Hypotheische Skizze des monophyletischen Ursprungs und der Verbreitung der 12 Menschen-Species von Lemurien aus üe r die Erde*. Although modified in subsequent editions, it is the only plate that survives through to at least the 9th edition (Table 1.2).

For the English translation of Natürliche Schöpfungsgeschichte, The History of Creation, or the Development of the Earth and Its Inhabitants By the Action of Natural Causes ("… or, as Professor Haeckel admits it would have been better to call his work, 'The History of the Development or Evolution of Nature,'"

---

* I have been able to examine only the 1st, 2nd, 5th, 7th, 8th, and 9th editions.

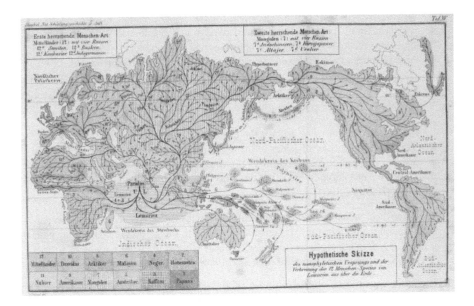

FIGURE 1.7 Reproduction of *Hypotheische Skizze des monophyletischen Ursprungs und der Verbreitung der 12 Menschen-Species von Lemurien aus tie r die Erde* from the 8th edition of *Natü liche Schöpfungs geschichte* (Haeckel, 1870, Taf. XV).

Huxley, 1869, p. 3), the 1st — 3rd editions (adapted and translated from the 2nd –7th German editions) have just four plates (see above) but many synoptic tables (Table 1.2). Among those four plates is the *Hypothetical Sketch of the Monophyletic Origin and the Extension of the 12 Races of Man from Lemuria over the Earth** (Haeckel, 1876a, Taf. XV, Figure 1.7).

Systematische Phylogenie (1894–1896) was Haeckel's final attempt to document the genealogical relationships for all of Life, some 26 years after his *Generelle Morphologie*. *Systematische Phylogenie* did indeed contain many pedigrees, but the 'oak-trees' of the *Generelle Morphologie* had disappeared, leaving just a series of stylised line drawings, similar to those from later editions of *Natürliche Schöpfungs- geschichte,* where taxa either link directly to each other, descending from one another. Each *Stammbaum* accompanied by a *System* (Table 1.2).

Toward the end of his life, Haeckel returned to the theme of evidence derived from the "three great records" — his threefold parallelism, the data to support his pedigrees:

The first rough drafts of pedigrees that were published in the *Generelle Morphologie* have been improved time after time in the ten editions of my *Naturaliche Schopfungs-*

* Haeckel seems to have made an error on the figure legend, as the diagram is supposed to be of the 12 *species* of man, rather than races, of which Haeckel recognised 36 (Haeckel, 1876a, pp. 308–309). Comparison with the German editions suggests the error was in translation as the word Menschen-Species is used.

geschichte (1868–1902) (English translation; *The History of Creation,* London, 1876). A sounder basis for my phyletic hypotheses, derived from a discriminating combination of the three great records — morphology, ontogeny, and palaeontology — was provided in the three volumes of my *Systematische Phylogenie* (Berlin, 1894–96) (Haeckel, 1909).

Yet even while acknowledging the effica y of the 'threefold' evidence, he continued to rely heavily on palaeontology:

... The task of phylogeny is to trace the evolution of the organic stem or species — that is to say, of the chief divisions in the animal and plant world, which we describe as classes, orders, etc.; in other words, it traces the genealogy of species. It relies on the facts of palaeontology, and fills the gaps in this by comparative anatomy and ontogeny (Haeckel, 1904, p. 97).

The three most valuable sources of evidence in phylogeny are palaeontology, comparative anatomy, and ontogeny. Palaeontology seems to be the most reliable source, as it gives us tangible facts in the fossils which bear witness to the succession of species in the long history of organic life (Haeckel, 1904, p. 393).

Agassiz questioned the usefulness of palaeontology as a means for detecting genealogy:

The work of palaeontology that relates to the ideas of Darwin seems to me to make the same mistakes as early efforts in zoology. One recognizes the very extensive resemblances between some animals that lived in different eras; the resemblances are often stronger if the successive types are closer than others, in both time and space. But we may find some identical analogies with all eras of the history of the earth and the same facts are repeated until the present era. Therefore, there is no need to resort to a chronological element in order to explain the origin. Moreover, next to the series of similar forms, of which we recognize their existence, even in the Recent era, we have some isolated and independent types (Agassiz, 1869a, p. 375, translation modifie from http://www.athro.com/general/atrans.html).

Thus, Agassiz rejected palaeontology as a source of information that might enlighten discovery of the *origin* of species, or at least its role in establishing species 'pedigrees', simply because "we have some isolated and independent types". Agassiz refers, almost indirectly, to his views on the origin of species, with the possibility of many independent origins giving rise to the same species in different places — a polygenic theory, relating to disjunct distributions (see below).

Haeckel's interest in 'creating' ancestors was hindered by the lack of paleontological 'facts', but that did not stop him from creating "missing ancestors modelled on living embryos" (Desmond, 1994, p. 349), or from the paraphyletic non-groups of systematics. The reliance on palaeontology, of course, set the scene for nearly all phylogenetic research during the following 100 years, a programme that eventually split at its seams once the concept of relationship was clarified and understood (Hennig, 1966) and the cladistic revolution could set about reforming palaeontology (Nelson, 1969; Patterson 1977) and, as a consequence, comparative biology (Williams & Ebach, 2004). If certain aspects of early 20th century German morphology had been acknowledged and understood, the revolution in palaeontology might not have been necessary (Williams & Ebach, 2004). Nevertheless, the majority of Haeckel's genealogical trees were linear schemes of relationships, taxa 'giving rise' to other taxa and paraphyletic groups not so much created but explained in terms of evolutionary relationships, relative to particular model of change, models still invoked today to account for paraphyly (Mayr & Bock, 2002).

HAECKEL'S HYPOTHEISC HE SKIZZE DES MONOPHYLETISCHEN URSPRUNGS UND DER VERBREITUNG DER 12 MENSCHEN-SPECIES VON LEMURIEN AUS ÏER DIE ERDE' AND THE CONCEPT OF CHOROLOGY

Of all Haeckel's genealogical diagrams, the most reproduced is that which firs appeared in *Anthropogenie,* Haeckel's popular book on the evolution of man, the tree Oppenheimer identified as *Quercus robur,* the European Oak (Oppenheimer, 1987, p. 127), the illustration still reproduced today, often for the cover illustration for books dealing with some aspect of Darwin's thinking, rather than Haeckel's (e.g., Richards, 1987; Bowler, 1988 [1992]; Alter, 1999). Its popularity may be because of its depiction of the evolution of man, possibly the first of its kind. The 1st German edition of *Anthropogenie* (1874b) includes fi e 'pedigrees' in all, but this particular tree guides the reader from the bottom to the top, from the primitive and insignifican Monads to the crowning glory of Man (Ruse, 1997), successively leading from one to the other up the sturdy trunk, with the rest of 'creation' splitting off at various intervals, leaving their ancestors in its wake (Figure 1.2b; Haeckel, 1874b, Taf. XII; 1883, Taf. XV, 1891, Taf. X;* the translated English edition, *The Evolution of Man: a Popular Exposition of the Principal Points of Human Ontogeny and Phylogeny,* first published in 1879b, includes the tree, as do all subsequent editions).

Less well known is Haeckel's *Hypotheische Skizze des monophyletischen Ursprungs und der Verbreitung der 12 Menschen-Species von Lemurien aus über die Erde or Hypothetical Sketch of the Monophyletic Origin and the Extension of the 12 Races of Man from Lemuria over the Earth* first published in Natürliche Schöpfungsgeschichte (Figure 1.7, Haeckel, 1870, Taf. XV). The implications behind the diagram are many but have been discussed by few (Nelson, 1983; Kirchengast, 1998; Nelson & Ladiges, 2001); this tree appeared as a dust jacket illustration (Nelson & Platnick, 1981).

The *Hypotheische Skizze* is a departure from those Haeckel previously drew, as it depicts the 'species' of man overlain on a map of the world, tracing out the

---

* The number of illustrations in *Anthropogenie* changed with successive editions, like many of Haeckel's books. The 1891 4th German edition, for example, has two 'oak tree' diagrams, one modified from previous editions (Taf. XV, which shifts the 'Reptilien' from branch to trunk, descendant to ancestor), the other illustrating the "Palaeontologischer Stammbaum der Wirblthiere" (Haeckel, 1891, Taf. XVI). Later English editions (the 1905 edition, for example) lack the familiar oak-tree diagram of Taf. XV (Figure 1.2b). A manuscript version of the tree depicted in Taf. XV is reproduced in Klemm (1969, pp. 112–13, as well as part of the dust jacket illustration), Gasman, 1971 [2004], pp. 8–9) and on the dust jacket and cover of Richards, (1987 [1989]).

migration route each human 'species' (= branch) took, arriving at its current place of habitation. The *Hypotheische Skizze* went through some changes with successive editions of *Natürliche Schöpfungsgeschichte,* the first version appearing in the 2nd edition (Haeckel, 1870, Pl. XV) as a half-tone, with the 'origin' of man located somewhere between Malay and South West Africa, a place he referred to as 'Paradise,' a point located out to sea. This particular viewpoint remained until the 7th edition, although the shading of the picture was somewhat enhanced (Haeckel, 1879a, Pl. XV). By the 8th edition this illustration was the only genealogy included as a plate (rather than a diagram in the text), had been rendered in colour, and 'Paradise' had moved onto land (Figure 1.8; Haeckel, 1889, Taf. 20; see Kirchengast, 1998, 178, Abb. 2). For the first English edition, the illustration was in colour, but 'Paradise' remained out at sea (Figure 1.9; Haeckel, 1876a, Taf. XV). In each case, when the reader encounters in the text the table for *Menschen-Arten und Rassen,* they are referred to this diagram*.

The notion of humans travelling from their 'original point of creation' has probably caused vast problems, not least in how we view ourselves (Bowler, 1995). Nevertheless, that they travelled retained a special place in the interpretation of current distributions. William Diller Matthew, for example, in his *Climate and Evolution* (1915) — written while Haeckel was still alive — also drew a map depicting the various routes of humans, stating that most "authorities … today agreed in placing the center of dispersal of the human race in Asia" (Matthew, 1915, p. 41; Figure 1.6; reproduced in Lomolino *et al.* 2004, p. 244). The route maps continue, seen in many popular books and articles on human evolution and migration (Finlayson, 2005).

The significance of the *Hypotheische Skizze,* human wanderings to one side, is that Haeckel chose it as a graphic representation of *chorology,* a term first proposed in his *Generelle Morphologie* (1868, pp. 286–289: "… chorology is the science of the geographic and topographic spread of organisms", translated), and expanded into a full chapter for *Natü liche Schpfungs geschichte* (*History of Creation*):

> I mean *Chorology,* or the theory of the *local distribution of organisms over the surface of the earth.* By this I do not only mean the *geographical* distribution of animal and vegetable species over the different parts and provinces of the earth, over continents and islands, seas, and rivers, but also their *topographical* distribution in a *vertical* direction, their ascending to the heights of mountains, and their descending into the depths of the ocean (Haeckel, 1925, p. 364).

Haeckel goes on:

> The strange chorological series of phenomena which show the horizontal distribution of organisms over parts of the earth, and their vertical distribution in heights and depths, have long excited general interest. In recent times Alexander Humbolt and Frederick Schouw have especially discussed the geography of plants, and Berghaus, Schmarda, and Wallace the geography of animals, on a large scale … only since Darwin that we have been able to speak of an independent science of Chorology … (Haeckel, 1925, pp. 365–366**).

---

* In each case the orgin of man was Lemuria.
** The various editions of *The History of Creation* have different wordings, but their essence is the same. This passage and further parts of Chapter XIV are the version reproduced in Lomolino et al., 2004, pp. 178–193.

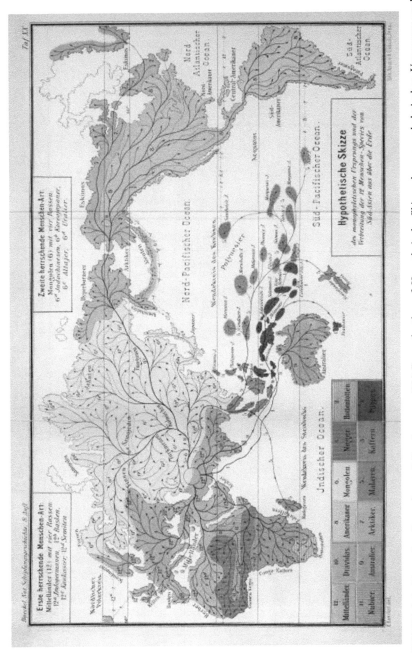

FIGURE 1.8 (See color figure insert following page 76.) Reproduction of *Hypotheische Skizze des monophyletischen Ursprungs und der Verbreitung der 12 Menschen-Species von Lemurien aus über die Erde* from the 8th edition of *Natürliche Schöpfungsgeschichte* (Haeckel, 1889, Taf. 20).

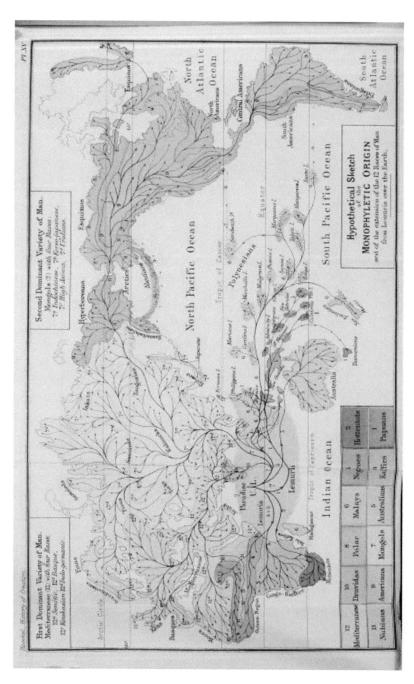

FIGURE 1.9 (See color figure insert following page 76.) Reproduction of *Hypothetical Sketch of the Monophyletic Origin and the Extension of the 12 Races of Man from Lemuria over the Earth* (Haeckel, 1876a, Taf. XV).

Haeckel states some conditions:

The most important principle from which we must start in chorology, and of the truth
on which we are convinced by due examination of the theory of selection, is that, as
a rule, every species has arisen only *once* in the course of time and only in *one* place
on the earth — its so-called 'centre of creation' — by natural selection ... the distribution of the great majority of animals and vegetable species in regard to which the
single origin of every species in a single locality, in its so-called 'central point of
creation,' can be considered as tolerably certain (Haeckel, 1925, p. 367).

Haeckel used the remaining parts of the chapter to discuss the various means of
migration species may undergo to travel from their particular centre of creation. He
used the idea sparingly in his own work, discussing the vertical and horizontal
distribution of species of *Radiolaria* found in the Challenger material (Haeckel,
1887a) and Australian material he later studied (Haeckel, 1893*).

Haeckel understood chorology as part of Physiology (= "The Science of Func-
tions"), noting that it was the "science of migrations" (Haeckel, 1904, pp. 98, Third
Table). Physiology (= "The Science of Functions") was contrasted with Morphology
(= "The Science of Forms").

THE DEVELOPMENT OF CHOROLOGY

According to Uschmann (1972), Haeckel's insistence on the fundamental importance
of Darwin's ideas to biology is most apparent in his 'ecology' and 'chorology',
"both concepts have won acceptance" (Uschmann, 1972), while for Hoßfeld chorology has "been widely adopted" (Stauffer, 1957; Hoßfeld, 2004, p. 84). Interestingly enough, in the various editions of *The History of Creation,* Haeckel appears
to contrast regional biogeography (of which more below) — exemplified by the
works of Humboldt, Schouw, Berghaus, Schmarda, and Wallace — with chorology.
Regional biogeography gained many early critics; writing while Haeckel was still
alive, Ortmann suggested:

It is incorrect to regard the creation of a scheme [of regions] of animal distribution as
an important feature or purpose of zoogeographical research. Thus we are justified in
saying that zoogeographical study, as introduced by Wallace [and Sclater], is not
directed in the proper channels [and results in] fruitless discussions on the limits of
the zoogeographical regions (Ortmann, 1902a, after Heads, 2005c, p. 87).

Ortmann understood the matter as one of linking the present to the past, along
with any necessary changes in the Earth's surface that might have occurred (such a
view has re-surfaced, Donoghue & Moore, 2003). Ortmann later, reviewing the work
of Jacobi (1900), noted that he (Jacobi) found 'certain parts of the Earth's surface'
that are 'inexplicable by the present conditions' (O[rtmann], 1902b, p. 158).

---

* Haeckel included a table of relationships from Haeckel 1866 (on p.xii) and tables of 'phylogenetische
stufenreihe' (on p. xiv).

With such criticisms, a distinction began to develop between what was eventually described as the static geographical method (regional biogeography) and the dynamic faunal method (chorology), summarised later by Voous:

> The geographical method is static; it tries to define the borders of zoogeographical regions, districts, or provinces. It is part of the classical zoogeography of Philip Lutley Sclater and Alfred Russel Wallace. The faunal method is dynamic; it tries to detect and to describe the far-reaching intergradation of separate faunas throughout the continents … . This method starts from the conception that there are distinct faunas but no distinct zoogeographic regions (Voous, 1963, p. 1104).

In an early paper, Ernst Mayr made the following comments:

> Eventually it was realised that the whole method of approach — the *Fragestellung* — of this essentially static zoogeography was wrong. Instead of thinking of fi ed regions, it is necessary to think of fluid faunas … (Mayr, 1946, p. 5; Mayr, 1976 [1997], p. 567).

Later he commented on regions and faunas:

> I shall not rehearse the history of zoogeography in the last 100 years. I shall merely remark that the faunal and historical approach favored by Darwin tended to recede into the background as the geographical approach of Sclater and Wallace came to the fore and as an increasing number of authors expended their energies in trying to determine the borders between geographic regions and in subdividing these regions into subregions and biotic provinces … (Mayr, 1965, p. 474, Mayr, 1976 [1997], p. 553).

Mayr continues by noting those who supported the 'faunal' approach, citing a paper by Carpenter (1894), an early critic of regions, noting a paper by Dunn, who "… was the pioneer of this concept [dynamic faunas]" (Dunn 1922), finishing with two examples from German ornithologists, Stegmann (1938) and Stresemann (1939) (Mayr, 1946, p. 5; Mayr, 1976 [1997], pp. 567–568). Yet, the faunal approach reaches further back than Dunn or German ornithology, back to Haeckel and his chorology. Mayr stated the difference between classical and 'modern' biogeographers:

> Classical zoogeography asked: What are the zoogeographic regions of the earth, and what animals are found in each region? The modern zoogeographer asks when and how a given fauna reached its present range and where it originally came from; that is, he is interested in faunas rather than in regions (Mayr, 1946, p. 6; Mayr, 1976 [1997], p. 569).

In Mayr's 'classical' zoogeography, his words clearly reflect not a static approach to the living world but a classificatory one: "to determine the borders between geographic regions and in subdividing these regions into subregions and biotic provinces …". The modern approach, as he outlined it, is more concerned with mechanisms and dealing with each 'lineage' separately, as if they developed independently of each other.

In the immediate past many have attempted to distinguish between different kinds of biogeography, such as ecological and historical biogeography (Humphries & Parenti, 1999). More recently biogeographers have suggested more subdivisions and more partitions — and, as a consequence, there is less coherence rather than clarity (Crisci et al., 2000, 2003; see also Morrone, 2005a, Table 1, correction in Morrone, 2005b, p. 1505). Given the above, notwithstanding various claims to the contrary, there is a clearer way of distinguishing the subject matter of the geographical distribution of organisms:

Biogeography is the study of the interrelationships of areas (classification);

Chorology is the study of the mechanisms of distribution related to taxon origins*.

If it is accepted that biogeography is the study of the interrelationships of areas and is a problem of classification, then its most significant aspect is the concept of homology, the parameter that allows hypotheses concerning relationships (Morrone, 2004, see below).

For chorology, mechanisms encompass all those suggested above, including all aspects of dispersal and vicariance, and whatever other theories relate to attempts to discover any particular taxon's origin (see another definition in Aubréville, 1970, p. 450). It is not necessary to trace the fortunes of chorology and its changing role, or even to ascertain whether it really did achieve general acceptance — the word is still in use, if not that often and in quite a variety of different ways (e.g., Huxley et al., 1998), but many 'modern' studies do appear to be within that remit (De Queiroz, 2005):

This new view [more oceanic dispersal] implies that biotas are more dynamic and have more recent origins than had been thought previously ... . The new support for oceanic dispersal has come primarily from information on the timing of speciation, fueled by the development of improved methods of DNA sequencing and of estimating lineage divergence dates based on molecular sequences (De Queiroz, 2005, p. 69).

De Queiroz's claim seems too bold. Support for timing of speciation and estimating lineage divergence dates come from palaeontology, fossils, Haeckel's 'significant data. Nevertheless, De Queiroz essentially embraces what is chorology.

It is worth noting that Erwin Stresemann (1889–1972, Haffer et al., 2000), noted above as an early proponent of dynamic faunal studies, was professor to a young Ernst Mayr, who saw the development of chorology within the German ornithological community of the 1930s (Junker, 2003; Bock, 2004). According to Mayr "... Virtually everything in Mayr's 1942 book was somewhat based on Stresemann's earlier publications" (Mayr, 1999, p. 23, 1997). According to Bock, Mayr's "firs and last interest in ornithology is biogeography ...", where he applied "the (then) new ideas for analyzing the biogeography of birds that were advocated by Stresemann (1939)" (Bock, 1994, p. 291, 2004, p. 645). According to Haffer et al., "After his high school examinations (Abitur), Stresemann entered the University of Jena in 1908, where he took courses offered by Ernst Haeckel" (Haffer et al., 2000, pp. 399–400). In 1931, Stresemann "declared historical morphology as terminated, i.e.

---

* It is possible that chorology is the same as Cain's Areography (Cain, 1944), a term Hubbs noted was "of bastard origin and denotes a concept that is sufficiently covered by 'biogeography' and by 'chorology'" (Hubbs, 1945).

phylogenetic or systematic morphology … in the sense of Ernst Haeckel … and Max Fürbringer …" (Haffer et al., 2000, 420).

Stresemann was also professor to Wilhelm Meise (1901–2002) (Haffer, 2003, p. 117, Meise "was primarily a systematist studying problems of geographical variation, hybridisation, and speciation in birds"). Meise nurtured the 19-year-old soon to be entomologist, Willi Hennig (1913–1976) (Meise & Hennig, 1932, 1935; Schmitt, 2001; Haffer, 2003), who went on to create and develop the chorological method for determining character polarity from geographical distances, a fl wed method as it turned out*. Hennig noted "The use of the chorological method in zoology has become known particularly through the books of Rensch" (Kiriakoff, 1954; Hennig, 1966, p. 133, 1950, pp. 192–199); Rensch was also a student of Stresemann (Junker, 2003).

Simpson, seemingly, discovered chorological relationships independently, through the use of clines: "Clines may, then, be distinguished according to the variate that is used to define the array [of populations]. In one case the arrangement is geographical and these may be called chorocllines, i.e., 'space clines.'" (Simpson, 1943, p. 174).

Chorology — in all its variations and permutations — addresses issues that pertain to the origin of things, either taxa or characters, in their geographical dimension.

ORIGINS

The origin of taxa, if at all a concrete notion to pursue, lies beyond the empirical horizon of systematics (De Pinna, 1999, p. 363).

The origin of species (Darwin, 1859), as well as the origin of taxa (Løvtrup, 1987), is a concern that pre-dates Darwin's efforts at explaining the phenomenon. For example, when Arthur Lovejoy surveyed "The Argument for Organic Evolution Before *The Origin of Species*" (Lovejoy, 1909a, b), he summarised three possibilities for the origin of species, possibilities that were outlined in a popular, standard account, Gray and Adams' *Elements of Geology* (1852; see also Lyell, 1832): (1) successive special creations; (2) "transmutation, which supposes that beings of the most simple organization having somehow come into existence, the more complex and the higher orders of animals have originated in them by a gradual increase in the complexity of their structures; and (3) *generatio aequivoca* [spontaneous generation] of individuals and species" (Lovejoy, 1909a, p. 500). Gray and Adams opted for the first, successive special creations (the option favoured by Agassiz and Bronn). It is not difficult to imagine more or even endless possibilities for speculating on the origin of species. Nevertheless, it was clear that somehow the geographical distribution of organisms would, if not provide clues, then at least point the way (Hofsten, 1916). For various reasons, the single centre of origin became the focus. Darwin wrote, "… the simplicity of the view that each species was first produced within a single region captivates the mind" (Darwin, 1859, p. 352).

Exploring species' origin today is much more precise, requiring the identificatio of a particular centre of origin (Avise, 2000; in the past, centres of origin have

---

* Hennig also coined the words *apochor* and *plesiochor,* related to geographical criteria (Hennig, 1950).

fluctuated in size; see Croizat et al., 1974, p. 269, footnote 6), the particular time of origin (Donoghue & Smith, 2003), and subsequent migration to other parts of the world (De Quieroz, 2005), investigations once considered to be the domain of palaeontology (Matthew, 1915) but now more closely associated with the various methods subsumed under phylogeography (Excoffie , 2004).

This original idea was popularised by Linnaeus, having its own 'origin' in the Bible with the notion that each species would begin minimally with two individuals: "If we trace back the multiplication of all plants and animals ... we must stop at one original pair of each species". Darwin found the idea captivating:

> Undoubtedly there are many cases of extreme difficulty in understanding how the same species could possibly have migrated from some one point to the several distant and isolated points, where now found. Nevertheless the simplicity of the view that each species was first produced within a single region captivates the mind. He who rejects it, rejects the vera causa of ordinary generation with subsequent migration, and *calls in the agency of a miracle.* It is universally admitted, that in most cases the area inhabited by a species is continuous; and when a plant or animal inhabits two points so distant from each other, or with an interval of such a nature, that the space could not be easily passed over by migration, the fact is given as something remarkable and exceptional (Darwin, 1859, p. 352).

However captivating Darwin's view might have been, it was questioned, primarily but not exclusively, by vicariance biogeographers (Croizat et al., 1974, p. 274; Croizat, 1981, p. 503, 1984, pp. 57–58; Patterson, 1983, p. 14; Heads, 1985, 2005a, p. 676, 2005b, p. 74, 2005c, p. 85, 94, 111; Brady, 1989, p. 112; Nelson & Ladiges, 2001, p. 401). The American ecologist Frederick Clements, early in the 20th century commented on Darwin's "captivating idea", noting that

> From the very nature of his task, Darwin was forced to assume that species were firs produced at one spot ... . This view seems to be little more than inheritance from the special creationists (Clements, 1909, p. 145).

In his review of the 1979 American Museum of Natural History's vicariance biogeography symposium (Nelson & Rosen, 1981), Ernst Mayr (1982a, p. 619) commented on the various critiques of the single point of origin, noting that "... Darwin's argument was directed against authors like Louis Agassiz, who explained discontinuous distribution of species by multiple independent creations." Darwin did not explicitly mention Agassiz. Nevertheless, Agassiz had indeed written on the subject of centres of origin:

> The greatest obstacles in the way of investigating the laws of the distribution of organized beings over the surface of our globe, are to be traced to the views generally entertained about their origin. There is a prevailing opinion, which ascribes to all living beings upon earth one common centre of origin, from which it is supposed they, in the course of time, spread over wider and wider areas, till they finally came into their present state of distribution. And what gives this view a higher recommendation in the opinion of most men is the circumstance, that such a method of distribution is considered as revealed in our sacred writings (Agassiz, 1850a, p. 181).

Agassiz, like those before, traced the idea of centres of origin to the Bible, and because he thought otherwise, saw himself in conflict with those "sacred writings", whereas for Darwin, those who disbelieve the "ordinary generation [from a single region] with subsequent migration" require "the agency of a miracle".

Sacred texts and miracles to one side, Leon Croizat (1894–1982, see Nelson, 1973; Craw, 1984a, b; Zunino, 1992; Llorente et al., 2000; Morrone, 2000; Colacino & Grehan, 2003), for example, happily rejected Darwin's "captivating idea". In more recent times Croizat's ideas have had significant representation by Michael Heads, among others (Grehan & Ainsworth, 1985), and have been related to what he (Heads) called the "nature of ancestors" (Heads, 1985). The notion that characters of ancestors might not be 'uniform' relates to Croizat's *polytopism* (= "multiple origins", Heads, 1985, p. 209; see Croizat, 1971, and Aubréville, 1969, 1974, 1975a, b), as contrasted by the more usual *monotopic* ancestor, a position Heads (and Croizat) suggested was held by the cladists Willi Hennig and Lars Brundin (Heads, 1985). Croizat favoured a polytopic origin for many taxa (Croizat, 1971, 1978) rather than the more usual monotopic explanation. Theodor Just, a palaeobotanist, offered the following commentary on polytopic speciation in review:

> This author [Suessenguth 1938] is of the opinion that certain areas can be explained best by assuming the possibility of a polytopic as well as polyphyletic origin, provided the original stock was sufficiently widely distributed. According to this view, several or even many parallel lines are evolving more or less simultaneously in several species and genera. This mode of origin would do away with the problem of large scale migrations and the invariably long spans of time required for such (Just, 1947, p. 132).

Just continues that "… not a single case is presented by Suessenguth to illustrate this view". But such options were — and still are (Heads, 1985) — considered, with vicariance as a covering explanation. Just's account to one side, the idea does not appear too dissimilar to that proposed by Agassiz (whom Croizat does not cite), if primary causes (deities) are disregarded.

This difference — between 'polytopism' and 'monotopism' — might go some way toward explaining the antagonism between Croizat and the Hennigians, at least with respect to different approaches for the 'origin' of taxa (Croizat, 1982). If differences of opinion are related to particular 'kinds' of origin and these differences in turn relate to particular mechanisms, then one might conceivably see the problem residing, once again, in a model — 'polytopism' or 'monotopism' — and its explanation — dispersal or vicariance. Oddly enough, given the contrast between Agassiz and Darwin and Croizat and Hennig, for example, evolution in its most general sense, seems not to be an issue — it is the idea that one might discover something about a taxon's *origin*.

Agassiz was specific enough concerning the role of geographic distribution:

> … work … in the Museum, has already extended to comparisons … with the view of ascertaining whether there is any probability of tracing a genetic connection between the animals of … different geographical areas, and how far geographical distribution and specific distinction are primary factors in the plan of creation. It must be obvious that the question of the origin of species is not likely to be discussed successfully

before the laws of geographical distribution of organized beings have been satisfactorily ascertained (Agassiz, 1865, p. 12; Annual report for 1864, cited in Winsor, 1991, p. 81).

His remarks echo those of Candolle's some 45 years earlier:

All of the theory of geographical botany rests on the particular idea one holds about the origin of living things and the permanence of species (Candolle, 1820, p. 417, translation in Nelson, 1978b, p. 285).

And Candolle's remarks relate to Lyell's discourse on the origin of species in the second volume of the *Principles of Geology* (Lyell, 1832; Nelson, 1978b), summaries of which are found in popular books like Gray and Adams' *Elements of Geology* above.

When Kinch, in his summary of ideas on the origin of life and the history of biogeography, noted in closing that "The key needed to resolve the biogeographical debate was a credible theory for species origin" (Kinch, 1980, p. 119), he appears mistaken. A credible theory, such models and their explanation, are best set to one side: "The origin of taxa, if at all a concrete notion to pursue, lies beyond the empirical horizon of systematics" (De Pinna, 1999, p. 363). If chorology is about the mechanisms to discover the *origin* of species, and the study of origins is a futile enterprise, then what is left?

As in systematics, there is classification (Nelson & Platnick, 1984). And without a classification of areas, what generalities, really, are there to explain?

REALMS, REGIONS, AND PROVINCES

Many early studies on biogeographic regions focused on humans, their place of residence and how they came to be where they are (Richardson, 1981; Browne, 1983). Human distribution and evolution were of some significance for Darwin, Desmond and Moore noting it in the *Origin,* if not in intent then in the text: "the subject pervades the text as a ghostly presence ..." (Moore & Desmond, 2004, p. xiv; see also Cooke, 1990).

While most acknowledge Buffon's *Histoire Naturelle* as the first real statements on the geographic distribution of organisms (Nelson, 1978b; Browne, 1983), it was Eberhard August Wilhelm von Zimmermann's (1743–1815) publication *Geographische Geschichte des Menschen, und der allgemein verbreiteten vierfüssigen Thiere, nebst einer hieher gehö igen zoologischen Weltcharte* that considered man as part of that distribution. Zimmermann dealt with the relation of the various 'kinds' of humans with their domesticated animals (Zimmermann, 1778). Zimmermann's work was not really well known; hence James Cowles Prichard (1786–1849) and his *Researches into the Physical History of Mankind* (1813) is often considered the first work to deal with biogeographic regions in relation to man as well as other animals. Prichard considered what areas were repopulated after the flood

... we may divide the earth into a certain number of regions, fitted to become the abodes of particular groupes of animals; and we shall find on inquiry, that each of these provinces, thus conjecturally marked out, is actually inhabited by a distinct nation of quadrupeds [mammals], if we may use that term (Prichard, 1826, p. 54, see Nelson & Platnick, 1981, p. 518; Kinch, 1980, p. 101).

TABLE 1.3
The Seven Regions Recognised by Prichard (1813, I, p. 53)

The Arctic Region of the New and the Old World
The Temperate
The Equatorial or Tropical
The Indian Islands
The Islands of New Guinea, New Britain and New Ireland, and those more remote in the Pacific
  Ocean
Australia
The Southern extremities of America and Africa

Prichard published several editions of his *Researches into the Physical History of Mankind** (Augustein, 1999), a work described by Moore & Desmond as "a monogenist encyclopedia" (Moore & Desmond, 2004), noting that while Prichard "defended Adamic unity; its arguments rested on a raft of biological, philological and enthnographic fact" (Moore & Desmond, 2004, p. xxvii). Prichard also included six maps illustrating "The natural history of man" depicting the areas occupied by various tribes and ethnic groups. Prichard named seven regions (Table 1.3). As "a monogenist encyclopedia", the book championed the view that the various races of man had one origin — were monogenic — and were one species, varieties being the result of environmental effects on waves of migration of humans — that is humans originated in one place and migrated to other areas (Haller, 1970).

One of the first books to deal with the entire animal kingdom and its geographical distribution was William Swainson's *A Treatise on the Geography and Classification of Animals* (1835). Swainson, elaborating on Prichard's work, began his commentary with some objections:

> The objections that may be stated against these [Prichard's] divisions chiefly arise from the author not having kept in view the difference between affinity and analogy, as more particularly understood by modern naturalists (Swainson, 1835, p. 13).

Swainson's comment is perhaps the first explicit statement relating to geographical homology, even though it is stated in the vernacular of the 19th century, as affinities and analogies. Swainson explained:

> The arctic regions of America, Europe, and Asia, indisputably possess the same genera, and in very many instances the same species; and if it should subsequently appear that these regions are sufficiently important in themselves to constitute a zoological province, then it is a perfectly naturally one; for not only are the same groups, but even the same species, in several instances, common to both. But can this be said of the second of these provinces,

---

* The second edition of *Researches into the Physical History of Mankind* (1826) is referred to rather than the 1st (1813) or later editions. The 1st edition is less explicit on details, while all later editions were influenced by Lyell (1830–1832). The 3rd edition consisted of fi e volumes, published between 1836 and 1847. The book was later renamed *Researches in the History of Mankind,* and later still *Researches in the History of Man,* with *The Natural History of Man* (1843) published as an abbreviated edition. An edition was reprinted in 1973, with an introduction written by G.W. Stocking (1973).

made to include the temperate regions of three continents? Certainly not. We find, indeed, analogies without end, between their respective groups of animals, but they have each a vast number of peculiar genera; and so few are the species common to all three, that the proportion is not perhaps greater than as 1 to 50 (Swainson, 1835, p. 13).

Between Prichard and Swainson three significant issues are bought into focus: the notion of distinct geographical regions, the notion of geographical homology, and the notion of the origin of things. Swainson went on to suggest that 'natural' geographical regions might very well correspond with "the fi e recorded varieties of the human species" (Swainson, 1835, p. 14).

For Agassiz, human distribution — as well as the distribution of other organisms — could be captured and understood in a series of realms or *Natural Provinces of Mankind,* areas which harbour collections of organisms, unique to each region (Figure 1.10 after Agassiz, 1854, facing p. lxxviii); for Haeckel, human distribution — as well as that of other organisms — was not so much captured by their current place of residence but by their travels (Figures 1.7–1.9).

AGASSIZS (1854) GEOGRAPHICAL REALMS: THE NATURAL PROVINCES OF MANKIND

In the recent compendium of 'classic' papers *The Foundations of Biogeography* (Lomolino et al., 2004), Louis Agassiz hardly rates a mention in all its 1400 pages. No contribution of his is included and, as far as can be established, he is referred to twice and then only in passing.* Agassiz's words may simply be of no significanc or importance today and not part of the inexorable fl w of common scientifi understanding. Ernst Mayr, never shy of an opinion, offered the following retrospective on Agassiz's geographical writings:

> When Agassiz, in the 1850s, wrote about biogeography, his uncompromisingly funda-
> mentalist interpretation seemed like a throwback to a long past period (Mayr, 1982b,
> p. 443).

That view to one side, Agassiz did have a lot to say about the geographical distribution of organisms, particularly in relation to man (Hofsten, 1916, pp. 297–301; Kinch, 1980, p. 102). Agassiz's interest in the geographical distribution of animals and plants began in 1845 (Agassiz, 1845a, b, 1846), forming a major part of his ideas on how a museum display should be constructed (Winsor, 1991, 2000).

Rather than simply a throwback, it is possible his contributions to geographical distribution were too closely linked with his shameful views on race and its developing context in mid-eighteenth-century United States (Roberts, 1982). Agassiz's

---

* The first mention of Agassiz's name is in Sclater's influential 1858 paper (Sclater, 1858, paper 9 in Lomolino et al., 2004, p. 131), where he discusses Agassiz's regions (see p. 10 of this paper); the second mention is in Clinton Hart Merriam's paper on biogeographical regions in the U.S. (Merriam & Stejneger, 1890, paper 15 in Lomolino et al., 2004, pp. 222, 228), where he mentions Agassiz's Great Central Province' (Lomolino et al., 2004, p. 222) and, in a footnote, Agassiz's *Louisana Fauna* (Lomilino et al., 2004, p. 228).

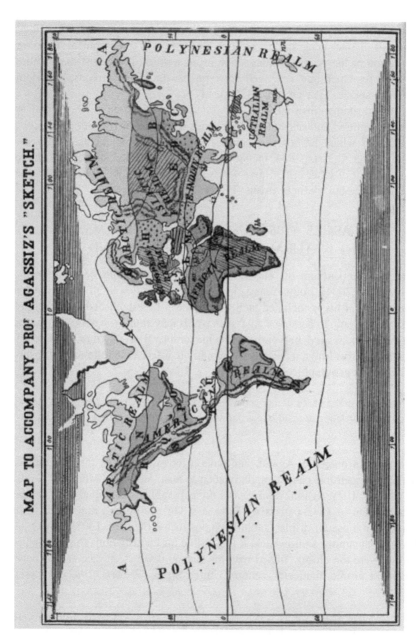

FIGURE 1.10 (See color figure insert following page 76.) Agassiz's map of realms harbouring collections of organisms unique to each region (Agassiz, 1854, facing p. lxxviii).

viewpoint has indeed fallen out of sight, except as an example used to illustrate the misfortunes that come of scientific ideas when given free social, religious or political reign and interpretation (see, for example, the different approaches taken to Agassiz and his views on 'Mankind' in Roberts, 1982, pp. 27–31; Stephens, 2000, pp. 195–211; Walls, 2003, pp. 181–184) — a fate Haeckel was to experience (Gasman, 1971 [2004], 1998). Nevertheless, beyond 'opinion' and 'context', Agassiz did have some interesting things to say; the story is worth telling from the geographical viewpoint. Agassiz published a short chapter outlining the zoogeographical regions of the Western hemisphere (Agassiz & Gould, 1848), followed by several papers dealing specifically with human distribution (Agassiz, 1850a, b, 1854). Agassiz's views on humans, both their origin and unity, changed between 1845 and 1855.

Before moving to his adopted home in the United States, Agassiz had lectured in Neuchâtel on the geographical distribution of organisms. It was the last in a series of 12 lectures on the 'Plan de la Création', the only one that was published (Agassiz, 1845a see also 1845b*). The series of lectures were highly publicised, and an advertisement from the time illustrates just how much Agassiz relied on the idea of a 'parallelism' to explain the 'Plan de la Création', these early words showing how he leaned more heavily on geography than he would come to do later:

> Suffice it to say that he [Agassiz] intends to show in the general development of the animal kingdom the existence of a definite preconceived plan, successively carried out; in other words, the manifestation of a higher thought, — the thought of God. This creative thought may be studied under three points of view: as shown in the relations which, in spite of their manifold diversity, connect all the species now living on the surface of the globe; in their geographical distribution; and in the succession of beings from primitive epochs until the present condition of things (Agassiz, 1885, p. 44).

It was in these 1845 Neuchâtel lectures that Agassiz first discussed his ideas of 'zoological provinces'. Agassiz noted that not only could provinces (or regions) be 'defined by their particular composition of plants and animals but also by their human inhabitants (Agassiz 1845a). Although Agassiz believed the animals confined to each region were all created within them — *in situ,* so to speak — he understood the races of Mankind to be "one and the same species capable of ranging over the surface of the globe" (Agassiz, 1845a, p. 29). Thus, his early views were distinctly monogenic. In spite of that sentiment, even in his early writings Agassiz had suggested that blacks ('Negroes') had a distinct origin, different from that of whites ('Caucasians'), and the former could not be traced back to the sons of Noah (Lurie, 1959, 1960). Suggesting that both had different origins, while forming part of his general opinion on the creation of all species, departed from the received religious view, of Man created as one. But at that time he did insist that all men belong in one species.

His Neuchâtel geography lecture was given in Boston shortly after Agassiz's arrival in the U.S. He travelled to Philadelphia, the place where he first met blacks and wrote the now infamous letter sent to his mother detailing his apparent distaste

---

* Agassiz dealt with geographical distributions in passing prior to 1845, notably in two papers from 1844 (Agassiz, 1844b, c) before his first general paper on animals and man (Agassiz, 1845a).

for these people*. Whether as a result of these early meetings or subsequent encounters, when Agassiz eventually lectured on man in Charleston he had changed his views, suggesting that black and white men were not only of different origin but of different species, a decidedly polygenic view.

In his chapter in the 1854 book *Types of Mankind,* Agassiz included a colour map to accompany his paper on 'Natural Provinces' (Figure 1.10; Agassiz, 1854, facing p. lxxviii). The map details the eight realms he recognised: Arctic, Asiatic, European, American, African, East-Indian (Malayan), Australian, and Polynesian. The realms and their inhabitants (human and other animals) are tabulated on a separate fold-out sheet (Agassiz, 1854, facing p. lviii), which has a page explaining the contents of the table (Agassiz 1854: lxxvii). The table's explanation expands on the composition of the realms, each being composed of a series of 'faunae' (except the Arctic). The table has eight columns and a variable number of rows; each row corresponds to a human 'kind' and a realm. In total, there are 67 numbered boxes. For example, column I is the Arctic realm, with 9 rows (numbered 1–9); column II is the Mongol realm, with 8 rows (numbered 10–17); column III is the 'European' realm, with 8 rows (numbered 18–25); and so on. Each column has a series of separate illustrations illustrating the 'concept' of each realm. For example, in the first box (number 18) of the European realm (the third column), there is an example of European man (in this case illustrated with a portrait of Cuvier [from his 1816 study], see Winsor, 1979, p. 113), followed by a human skull (box 19) and six other animals (a bear, a stag, an antelope, a goat, a sheep, and an aueroch [an ancient ox]). Only in the Arctic realm (column I, row 9) is there a plant — a reindeer-moss (a lichen). With respect to man, the realms marked on the coloured map are not equivalent to the entries in the table: the Arctic, European, American, Malay, and Australian are the same, whereas the Asiatic realm is "inhabited by Mongols ..." (column II in the table) and the African realm is "inhabited by Nubians, Abyssinians, Foolahs, Negroes, Hottentots, Bosjesmans"; the 'Negro' is the example used for column V (the African realm), and the Hottentot is used for the example of column VI, described in the legend of the table as the "Hottentot fauna". The Polynesian realm is not represented in the table at all, but the legend states "inhabited by South-Sea Islanders ..." (Agassiz, 1854, p. lxxvii).

For all their detail, Agassiz realms do not relate to each other. That is, Agassiz saw regions as problems of definition rather than of discovery — as in the discoveries made with the classifications of animals — in spite of the fact he had written a book on the subject (Agassiz, 1859).

REGIONS, HOMOLOGY, AND RELATIONSHIPS

The problem of classification was solved some years ago, or at least a solution was made possible. Appreciation of that solution has dwindled in recent years, simply because cladistics became confused with a particular method of analysis, as if it were one solution among many, as if in the geographical realm it was one aspect of chorological investigation. Cladistics is not a method nor a doctrine but a statement

---

* This letter has been reproduced on many occasions.

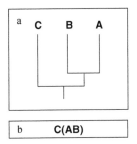

FIGURE 1.11 Cladistic parameter as a branching diagram (a) and a written statement (b).

about classification and its results, captured by what Nelson and Platnick called the cladistic parameter (Nelson, 1979, p. 12; Nelson & Platnick, 1981, p. 169 et seq.):

A and B are more closely related to each other than they are to C.

That statement can be represented as a diagram (Figure 1.11a) or as a simple notation (Figure 1.11b). Nevertheless, it might be difficult to imagine such statements of relationships residing in a matrix of binary characters, series of 0s and 1s, a largely phenetic construct, an issue that has been tackled from the perspective of systematics (Williams, 2004). The notion of homology in biogeography has been previously tackled, the first detailed statement being Patterson (1981b, p. 448, see Patterson, 1980, p. 238), which may be summarised as "... in cladistic biogeography homologies are congruent distributions of taxa ..." (Patterson, 1981b, p. 448, 466; Platnick & Nelson, 1989, p. 412; Nelson, 1994, p. 135); other interpretations are possible (Grehan, 1988; Morrone, 2001, 2004).

Morrone adopted the concepts of primary and secondary homology (De Pinna, 1991), relating each to particular methods of analysis (Morrone, 2001, 2004). Here I believe there is a mistake, one that currently pervades systematics. Primary homology is related to a phenetic view of similarity, somewhat divorced from any *direct* notion of relationship. Comparison among organisms reveal homologues, placing those homologues in context allows homology statements to be made — that is, statements of relationship derived from the data and independent of any particular method. In effect, there are no primary or even secondary homology statements; there are just statements of relationships (homology), some of which turn out to be true, others not. If biogeography concerns classification then it deals with relationships.

SCLATER, HUXLEY, AND THE CLASSIFICATION OF REGIONS

A paper of some significance in the study of regions is Philip Sclater's '*On the general geographical distribution of the members of the class Aves*' (Sclater, 1858). Sclater held the view that "each species must have been created within and over the geographical area, which it now occupies", a viewpoint not unlike that of Agassiz. Sclater commented on Agassiz's and Swainson's studies:

> In Mr. Swainson's article in Murray's 'Encyclopedia of Geography,' and in Agassiz's introduction to Nott and Gliddon's 'Types of Mankind,' what I consider to be a much more philosophical view of this subject is taken. The latter author, in particular, attempts to show that the principal divisions of the earth's surface, taking zoology for our guide, correspond in number and extent with the areas occupied by what Messrs. Nott and Gliddon consider to be the principal varieties of mankind. The argument to be deduced from this theory, if it could be satisfactorily established, would of course be very adverse to the idea of the original unity of the human race, which is still strongly supported by many Ethnologists in this country. But I suppose few philosophical zoologists, who have paid attention to the general laws of the distribution of organic life, would now-a-days deny that, as a general rule, every species of animal must have been created within and over the geographic area which it now occupies. Such being the case, if it can be shown that the areas occupied by the primary varieties of mankind correspond with the primary zoological provinces of the globe, it would be an inevitable deduction, that these varieties of Man had their origin in the different parts of the world where they are now found, and the awkward necessity of supposing the introduction of the red man into America by Behring's Straits, and of colonizing Polynesia by stray pairs of Malays f oating over the water like cocoa-nuts, and all similar hypotheses, would be avoided (Sclater, 1858, p. 131).

Sclater proposed six regions, Palaearctic, Ethiopian, Indian, Australian, Nearctic, and Neotropical, grouped into two series (Table 1.4). Huxley (1868) offered another version, retaining Sclater's six regions but grouping them differently, along a North-South divide (Table 1.4). Although Wallace adopted Sclater's regions, he prepared a diagram contrasting Sclater and Huxley's groupings

TABLE 1.4
Regions and their Relationships, after Sclater (1858) and Huxley (1868)

Sclater (1858)		Huxley (1968)	
Paleogaea		Arctogaea	
	Palaearctic		Palaearctic
	Ethiopian		Ethiopian
	Indian		Indian
	Australian		Nearctic
Neogaea		Notogaea	
	Nearctic		Australian
	Neotropical		Neotropical

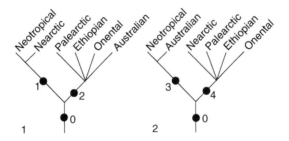

FIGURE 1.12 Contrasting regional classif cations of Huxley and Sclater as cladograms of areas (after Nelson & Platnick, 1981, Figure 8.50).

(Wallace, 1876, I, p. 66). Wallace's diagram is similar in some respects to Table 1.4. While the regions, Sclater proposed and Wallace endorsed were discussed almost endlessly in the following years, a signif cant change occurred some 120 years later. The contrasting classif cations of Huxley and Sclater were compared in Nelson and Platnick as two contrasting cladograms (Figure 1.12; after Nelson & Platnick, 1981, Figure 8.50). The cladogram representing Sclater's scheme (Figure 1.12, left) has two nodes corresponding to Paleogaea and Neogaea; the cladogram of Huxley's scheme (Figure 1.12, right) has two nodes corresponding to Arctogaea and Notogaea. In fact, one might say that the classif cations disagree over the placement of the Australian and Nearctic region. In any case, what remained was a problem in classif cation, not simply a static system that did not work or had no use. A suitable analogy would be if the classif cation of land plants was decided by general agreement, and then it was supposed to inform on 'greater' issues. Quite simply, classif cations of regions became of little use, simply because they were artif cial, adopted and agreed upon by f at. While the study of regions was understood as static, that characterisation seems deliberately obtuse. From this example, and representation of regions as cladograms, it is best understood as a problem of classif cation.

CROIZAT'S RADICAL REALMS: OCEAN BASIN AND CLADOGRAMS

Croizat presented a 'radical' (Craw & Page, 1988, Figure 12) re-arrangement of global realms (regions) (Croizat, 1958, Figure 259; see also Nelson & Ladiges, 2001, p. 393), focusing on f ve interrelated regions. Craw represented Croizat's diagram with two cladograms (Craw, 1983, Figure 4A, B, 1988, f g 13.11). He included three terminals, North America, New Zealand and Africa, and presented one diagram of their interrelationships, each terminal having more than one direct relationship (Craw, 1983, Figure 4A). The other diagram included reticulations, allowing all the terminals to appear only once, but the lines connecting them are duplicated (Craw, 1983, Figure 4B). The nodes of the diagrams were recognised as the Pacif c, Indian and Atlantic Oceans.

Nelson provided an alternative pair of cladograms to represent Croizat's realms. Nelson's f rst cladogram (re-drawn as Figure 1.13a) is unresolved, including just the

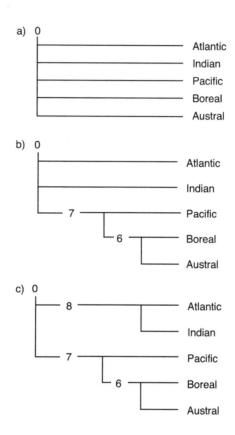

FIGURE 1.13 (a) Cladistic representation of Croizat's (1958, Figure 259) summary of global patterns of plants and animals, after Nelson (1985, Figure 1B). Node 0 = Gondwana. (b) Cladistic representation of Croizat's (1958, Figure 259) summary of global patterns of plants and animals, after Nelson (1985, Figure 1C; Nelson & Ladiges 2001, fig 8B). Node 0 = Gondwana, Node 6 = Bipolar, Node 7 = Pacific Rim. (c) Cladistic representation of Croizat's (1958, Figure 259) summary of global patterns of plants and animals, after Nelson & Ladiges (2001: Figure 8A). Node 8 = Gondwana, Node 6 = Bipolar, Node 7 = Pacific Rim.

five 'regions' Croizat recognised: Atlantic, Austral, Boreal, Indian and Pacific Nelson (1985, p. 188) said of these 'regions', that the Atlantic "includes taxa whose subtaxa are differentiated on either side of the tropical Atlantic", the Indian Ocean "similarly includes taxa having species endemic on or along the shores of the tropical Indian Ocean … So … for the tropical Pacific" The Boreal and Austral regions are somewhat different, and "include those taxa commonly termed Holarctic and Antarctic, and together termed bipolar or antitropical" (Nelson 1985, p. 188). This cladogram (Figure 1.13a) is uninformative of relationships, the implication being that Croizat represented (or discovered) the 'regions' but did not suggest how they might be further interrelated. Nelson's second cladogram (reproduced here as Figure 1.13b) presents two nodes worth of resolution, one representing 'bipolarity', the other the 'Pacific rim' (Nelson, p. 1985, Figure 1C; Nelson & Ladiges, 2001,

Figure 8B; see Brooks et al., 1981, Figure 16; Parenti, 1991, p. Figure 11; but see Lovejoy, 1996, 1997; De Carvalho *et. al.,* 2004).

Other representations of Croizat's diagram have been proposed, such as the series of blocks representing 'parts' of continents (Craw & Page, 1988, Figure 12; Craw, 1988, Figure 13.12, 1989, p. 537, Figure 8; Page, 1989, p. 475, Figure 5; Grehan, 1991, Figure 1; Craw et al., 1999, Figure 6–13; Humphries & Parenti, 1999, Figure 1.14; Grehan 2001; and in a modified form, Parenti, 1991, Figure 1; cf. Craw, 1982, p. 311, Figure 3). Nelson & Ladiges (2001, Figure 8A) include another cladogram, representing this novel proposition (Figure 1.13c).

In earlier times, regions were considered to be units in need of definitio (Schmidt, 1954; Horton, 1973) rather than discovery, as evidenced by the above series of hypotheses. Yet, interrelationships among regions, although not always clear, seem tractable (Cox, 2001; Morrone, 2002), even if competing systems require evaluation and testing. Ultimately, regions (and their subdivision) may come to represent the living world, in the same way angiosperms, vertebrates, and the many other taxa characterised over the years, are understood as part of the world we live in, discoveries, actual knowledge.

SUMMARY: THE THREEFOLD PARALLELISM: ... AND ITS END (NELSON, 1978A)

The focus of this paper was the threefold parallelism and what it meant in the hands of two very different people: Louis Agassiz and Ernst Haeckel. For Haeckel, it provided "the strongest proofs of the truth of the theory of evolution;" for Agassiz, the strongest evidence for the plan of divine creation. Haeckel developed his three-fold parallelism, hinging its usefulness on a version of the biogenetic law and the extent of data provided by palaeontology (Bryant, 1995). For Haeckel, geography was not considered direct evidence. Rather, it is circumstantial. Haeckel's efforts to forge a 'phylogenetic' system of representation did not go unchallenged, in the firs instance by a number of German morphologists, in the second by the cladistic revolution in palaeontology (Williams & Ebach, 2004).

Agassiz continued to require further sources of evidence, each drawing him nearer to the explanation of the origin of things, a world view where the primary cause was hidden from investigation (Rieppel, 1988; Janvier, 2003) — a cause Agassiz had already identified. The threefold parallelism came and went, eventually receiving its 'death-blow' at the hands of Nelson (1978a, although some still see value in its original formulation, Bryant, 1995 — although Bryant does note the geographical possibility: Bryant, 1995, p. 208). Briefl , Nelson concluded that the 'old' threefold parallelism (palaeontology, ontogeny, systematics) has no special meaning or interpretation beyond being interrelated evidence for the classification of organisms; biogeography, the spatial dimension, was the only independent source of data on organisms and their place in the world — and that too was a problem of classification (Nelson, 1978b). The issue of classification was solved, to a degree, with the clarification of relationships (Hennig, 1966), something only recently becoming apparent in the context of all kinds of data (Nelson, 1994). In a general sense, any parallelism of evidence is concerned with homology, or, perhaps, with a unit of classification

It might be said that a new threefold parallelism arose from the ashes of the old: Form, the interrelations of palaeontology, ontogeny, systematics — and now molecules, Space — the interrelations of areas (Ebach, 1999), and Time, falling out from their interplay (Nelson & Platnick, 1981) — time being a discovery rather than an imposition. This reformulation picks up a problem, f rst stated clearly by Candolle, developed by Prichard, Lyell, Swainson, Agassiz, Sclater (and others), eventually clarif ed by Nelson and Platnick (1981), who posed the problem in a tractable way. The revived threefold parallelism — systematics, biogeography, geology — does not aim at discovering ancestry, origins or any matters related to those things as commonly understood from Haeckel to our present state, but relates to the tractability of classif catory problems and what that allows us to discover of the world we live in.

ACKNOWLEDGEMENTS

I am grateful to Malte Ebach and Ray Tangney for their invitation to present this work at the 2005 biogeography symposium. Gary Nelson and Bob Press provided numerous comments on the text.

REFERENCES

Aescht, E. (1998). Ernst Haeckel — Ein Plädoyer für die wirbellosen Tiere und die biologische Systemtik. In *Welträ el und Lebenswunder. Ernst Haeckel — Wirk, Wirkung und Folgen. Stapfia* 56, N.F. 131, 19–83.

Agassiz, L. (1844a). *Recherches sur les Poissons Fossiles,* vol. 1.Text. Petitpierre, Neuchâtel.

Agassiz, L. (1844b). Sur la distribution géographique des quadrumanes. *Bulletin de la Société des Sciences Naturelles de Neuchâ l,* 1, 50–52.

Agassiz, L. (1844c). Sur la distribution géographique des Chiroptères. *Bulletin de la Société des Sciences Naturelles de Neuchâ l,* 1, 59–62.

Agassiz, L. (1845a). [Sur la distribution géographique des animaux et de l'homme]. *Bulletin de la Société des Sciences Naturelles de Neuchâtel,* 1, 162–166.

Agassiz, L. (1845b). Notice sur la géographique des animaux. *Revue Suisse,* 29–31.

Agassiz, L. (1846). [Observations sur la distribution géographique des êtres organisés]. *Bulletin de la Société des Sciences Naturelles de Neuchâ el,* 1, 357–362.

Agassiz, L. (1850a). Geographical distribution of animals. *Christian Examiner and Religious Miscellany* 48, 181–204 [*Bulletin de la Société des Sciences Naturelles de Neuchâ el,* 2, 347–350; *Edinburgh New Philosophical Journal,* 49, 1–25; *Verhandl. Naturhist. Ver. Preuss. Rheinland und Westphalens,* 7, 228–254].

Agassiz, L. (1850b). The natural relations between animals and the elements in which they live. *American Journal of Science,* 2nd ser. 9, 369–394 [*Annals and Magazine of Natural History,* 6, 153–179; *Edinburgh New Philosophical Journal,* 49, 193–227].

Agassiz, L. (1854). Sketch of the natural provinces of the animal world and their relations to the different types of Man. In *Types of Mankind: or, Ethnological Researches, Based upon the Ancient Monuments, Paintings, Sulptures, and Crania of Races, and upon their Natural, Geographical, Philological, and Biblical history: Illustrated by Selections from the Inedited [sic] Papers of Samuel George Morton, M.D.* (ed. by Nott, J.C. & Gliddon, G.R.; additional contributions from, L. Agassiz, W. Usher and H.S. Patterson), pp. lviii — lxxvii. Lippincott, Grambo & Co., Philadelphia.

Agassiz, L. (1857). Essay on Classif cation. *Contributions to the Natural History of the United States*. Vol. I. Little, Brown & Co., Boston, MA.

Agassiz, L. (1859). *An Essay on Classification*. London.

[Agassiz, L.] (1860–1862). [Obituary of Tiedemann]. *Proceedings of the American Academy of Arts and Sciences*, 5, 243–247.

Agassiz, L. (1862). Methods of study in natural history. *The Atlantic Monthly*, 9(51), 51–56. [Agassiz, L. (1863) *Methods of Study in Natural History*. Boston, MA].

Agassiz, L. (1865). *Annual Report of the Trustees of the Museum of Comparative Zoö ogy*. Harvard, Cambridge, MA.

Agassiz, L. (1869a) *De l'Espece et de la Classification en Zoologie*. Balliere, Paris, (English translation at http://www.athro.com/general/atrans.html).

Agassiz, L. (1869b). Principes rationnels de la classif cations Zoologique. *Revue des cours scientifiques de la France et de l'étranger*, 6, 146–155.

Agassiz, L. (1874). Evolution and permanence of type. *The Atlantic Monthly*, 1874, 92–101.

Agassiz, E.C.C. (1885). *L. Agassiz: His Life and Correspondence*. Houghton, Miff in, Boston, MA (reprinted by Thoemmes, Bristol, 2002).

Agassiz, L. & Gould, A.A. (1848). *Principles of Zoology*. Boston, MA.

Alter, S.G. (1999). *Darwinism and the linguistic image*. The John Hopkins University Press, Baltimore, MD.

Atkinson, Q.D. & Gray, R.D. (2005). Curious parallels and curious connections — Phylogenetic thinking in biology and historical linguistics. *Systematic Biology*, 54, 513–529.

Aubréville, A. (1969). A propos de "l'Introduction raisonnée a la biogéographie de l'Afrique" de Léon Croizat. *Adansonia*, 9, 489–496.

Aubréville, A. (1970). Vocabulaire de biogéographie appliquée aux régions tropicales. *Adansonia*, 10, 439–497.

Aubréville, A. (1974). Origins polytopiques des angiospermes tropicales: 2 [Polytopian origin of the angiosperms 2] *Adansonia*, 14, 145–198.

Aubréville, A. (1975a). The origin and history of the f oras of tropical Africa. Application of the theory of the polytopic origin of tropical angiosperms. *Adansonia*, 15, 31–56.

Aubréville, A. (1975b). Geophyletic studies on the Bombacaceae: The origin and history of the f oras of tropical Africa. Application of the theory of the polytopic origin of tropical angiosperms. *Adansonia*, 15, 57–64.

Augustein, H.F. (1999). Introduction. *James Cowles Prichard's Anthropology: Remaking the Science of Man in Early Nineteenth-Century Britain* (ed. by Augustein, H.F.), pp. ix–xx. Clio Medica S. Wellcome Institute Series in the History of Medicine, Editions Rodopi B.V., Amsterdam.

Avise, J.C. (2000). *Phylogeography: The History and Formation of Species*. Harvard University Press, Cambridge, MA.

Baez, A.M., Cione, A.L. & Torino, A. (1985). Diagramas ramif cados de relaciones en biologia comparada. *Ameghiniana*, 21, 319–331.

Baron, W. (1961). Zur Stellung von Heinrich Georg Bronn (1800–1862) in der Geschichte des Evolutionsgedankens. *Sudhoffs Archiv fü Geschichte der Medizin und der Naturwissenschaften*, 45, 97–100.

Bock, W.J. (1994). Ernst Mayr, naturalist: his contributions to systematics and evolution. *Biology and Philosophy*, 9, 267–327.

Bock, W.J. (2004). Ernst Mayr at 100: a life inside and outside ornithology. *Auk*, 121, 637–651.

Bouquet, M. (1995). Exhibiting knowledge: the trees of Dubois, Haeckel, Jesse and Rivers at the *Pithecanthropus* centennial exhibition. In *Shifting Contexts: Transformations in Anthropological Knowledge (ASA Decennial Conference Series: The Uses of Knowledge: Global & Local Relations)* (ed. by Strathern, M.), pp. 31–55. Taylor & Francis Books Ltd, London.

Bowler, P. (1976a). The changing meaning of 'evolution'. *Journal of the History of Ideas,* 36, 95–114.

Bowler, P. (1976b). *Fossils and progress: Paleontology and the idea of progressive evolution in the nineteenth century.* Science History, New York.

Bowler, P. (1988 [1992]). *The Non-Darwinian Revolution.* John Hopkins University Press, Baltimore, MD.

Bowler, P. (1995). The geography of extinction: biogeography and the expulsion of 'Ape Man' from human ancestry in the early twentieth century. In *Ape, Man, Apeman: Changing Views since 1600* (ed. by Corby, R. & Theunissen, B.), pp. 185–193. Department of Prehistory, Leiden.

Brace, C.L. (1981). Tales of the phylogenetic woods: The evolution and signif cance of evolutionary trees. *American Journal of Anthropology,* 56, 411–429.

Brady, R.H. (1989). The global patterns of life: A new empiricism in biogeography. *Gaia and Evolution,* pp. 111–126, Wadebridge Ecological Centre, Camelford, Cornwall, UK.

Bronn, H.G. (1858). *Untersuchungen üe r die Entwickelungs-Gesetze der organischen Welt.* E. Schweizerbart'sche Verlagshandlung, Stuttgart.

Bronn, H.G. (1859a). On the laws of evolution of the organic world during the formation of the crust of the earth. *Annals and Magazine of Natural History,* 3(4), 81–90, 175–184. (translation of Bronn, 1859b).

Bronn, H.G. (1859b). Recherche sur les lois d'evolution du monde geologique pendant la formation de la croute terrestriale. *Archives des Sciences Physiques et Naturelles,* 1859 (5), 217–241.

Bronn, H.G. (1860). Schlusswort des Übersetzers, in Charles Darwin, *Über die Entstehung der Arten im Thier- und Pflnz en-Reich durch natü liche Zä htung, oder Erhaltung der vervollkommneten Rassen in Kampfe ums Dasyn,* (based on 2nd English ed.), trans. H.G. Bronn, Schweizerbart'sche Verhandlung und Druckerei, Stuttgart.

Bronn, H.G. (1861). Essai d'une résponse à la question de prix proposée en 1850 par l'Academie des Sciences pour le concours de 1853, et pius remise pour celui de 1856, savoir: Etudier les lois de la distribution des corps organisés fossiles deans les differents terrains sédimentaires, suivant l'order de leur superposition. *Comptes Rendu Hebdomadaire des Séances de l'Académie des Sciences, Supplement,* 2, 377–918.

Bronn, H.G. (1863). Schlusswort des Übersetzers, in Charles Darwin *Über die Entstehung der Arten im Thier- und Pflnz en-Reich durch natü liche Zä htung, oder Erhaltung der vervollkommneten Rassen in Kampfe ums Dasyn,* 2nd ed. (based on 3rd English ed.), trans. H.G. Bronn, Schweizerbart'sche Verhandlung und Druckerei, Stuttgart.

Brooks, D.R., Thorson, T.B., & Mayes, M.A. (1981). Freshwater stingrays (Poamotrygonidae) and their helminth parasites: Testing hypotheses of evolution and coevolution. *Advances in Cladistics, The 1st meeting of the Willi Hennig Society* (ed. by Funk, V.A. & Brooks, D.R.), pp. 147–175. New York Botanical Gardens, New York.

Browne, J. (1983). *The secular ark: studies in the history of biogeography.* Yale University Press, New Haven, Connecticut.

Bryant, H.M. (1995). The threefold parallelism of Agassiz and Haeckel, and polarity determination in phylogenetic systematics. *Biology and Philosophy,* 10, 197–217.

Burkhardt, F, Porter, D.M., Browne, J. & Richmond, M. (1993). *The correspondence of Charles Darwin.* Vol. 8. 1860. Cambridge University Press, Cambridge.

Cain, S.A. (1944). *Foundations of Plant Geography.* Harper & Brothers, New York & London.

Candolle, A.P. de. (1820). *Essai Élémentaire de Géographie Botanique.* F.G. Levrault, Strasburg.

Carpenter, G. (1894). Nearctic or Sonoran? *Natural Science,* 5, 53–57.

Clements, F.E. (1909). Darwin's infuence upon plant geography and ecology. *American Naturalist,* 43, 143–151.

Colacino, C. & Grehan, J.R. (2003). Ostracismo alle frontiere della biologia evoluzionistica: il caso Léon Croizat. *Scienza e deomocrazia* (ed. by Capria, M.M.), pp. 195–200. Liguori, Napoli.

Cox, C.B. (2001). The biogeographical regions reconsidered. *Journal of Biogeography,* 28, 511–523.

Cooke, K.J. (1990). Darwin on man in the *Origin of Species. Journal of the History of Biology,* 26, 517–521.

Corsi, P. (1988). *The Age of Lamarck: Evolutionary Theories in France 1790–1830,* (trans. Jonathan Mandelbaum), University of California Press, Berkeley, California.

Cracraft, J. & Donoghue, M. (2004) *Assembling the tree of life.* Oxford University Press, New York, New York.

Craw, R.C. (1982). Phylogenetics, areas, geology and the biogeography of Croizat: a radical view. *Systematic Zoology,* 31, 304–316.

Craw, R.C. (1983). Panbiogeography and vicariance cladistics: are they truly different? *Systematic Zoology,* 32, 431–438.

Craw, R.C. (1984a). Never a serious scientist: the life of Leon Croizat. *Tuatara,* 27, 5–7.

Craw, R.C. (1984b). Leon Croizat's biogeographic work: a personal appreciation. *Tuatara,* 27, 8–13.

Craw, R.C. (1988). Panbiogeography: Method and synthesis in biogeography. *Analytical Biogeography* (ed. by Myers, A.A. & Giller, P.S.), pp. 405–435. Chapman & Hall, London.

Craw, R.C. (1989). New Zealand biogeography: a panbiogeographic approach. *New Zealand Journal of Zoology,* 16, 527–547.

Craw, R.C. (1992). Margins of cladistics: identity, differences and place in the emergence of phylogenetic systematics, 1864–1975. *Trees of Life: Essays in the Philosophy of Biology* (ed. by Griffths, P.), pp. 65–107. Kluwer Academic, Dordrecht.

Craw, R.C. & Page, R.D.M. (1988). Panbiogeography: method and metaphor in the new biogeography. *Evolutionary processes and metaphors* (ed. by Ho, M.-W & Fox, S.W.), pp. 163–189. John Wiley & Sons, London.

Craw, R.C., Grehan, J.R. & Heads, M.J. (1999). *Panbiogeography: tracking the history of life.* Oxford University Press, New York.

Crisci, J.V., Katinas, L. & Posadas, P. (2000). *Introducció a la teoría y prá tica de la biogeografá histó ica.* Sociedad Argentina de Botánica. Buenos Aires.

Crisci, J.V., Katinas, L. & Posadas, P. (2003). *Historical biogeography: an introduction.* Harvard University Press, Cambridge.

Croizat, L. (1958). *Panbiogeography or an introductory synthesis of zoogeography, phytogeography, and geology; with notes on evolution, systematics, ecology, anthropology, etc.* Vol. 1: The New World; Vol. 2: The Old World. (Bound as 3 vols.). i-xxxi, 2755 pp. Published by the Author, Caracas.

Croizat, L. (1971). Polytopisme ou monotopisme? Le cas de *Viola parvula* Tin. et de plusieurs autres plantes et animaux. *Boletinda da Sociedade Broteriana,* 45, 379–431.

Croizat, L. (1978). Hennig (1966) entre Rosa (1891) y Løvtrup (1977): medio siglo de "Sistemática Filogenética". *Boletim Academia de Ciencias Físicas, Matematicas y Naturales. Caracas,* 38, 59–147.

Croizat, L. (1981). Biogeography: past, present and future. *Vicariance Biogeography: a Critique* (ed. by Nelson, G. & Rosen, D.), pp. 510–523. Columbia University Press, New York.

Croizat, L. (1982). Vicariance/vicariism, panbiogeography, "vicariance biogeography," etc.: a clarifcation. *Systematic Zoology,* 31, 291–304.

Croizat, L. (1984). Mayr vs Croizat: Croizat vs Mayr — an enquiry. *Tuatara*, 27, 49–66.

Croizat, L., Nelson, G. & Rosen, D.E. (1974). Centers of origin and related concepts. *Systematic Zoology*, 23, 265–287.

Cuvier, G. (1816). *Le Règne Animal distribué d'après son organisation pour servir de base à l'histoire naturelle des animaux et d'introduction à l'anatomie comparée. Les reptiles, les poissons, les mollusques et les annélides*. Edn 1, Règne Animal, i–xviii + 1–532.

Cuvier, G. (1984). Biographical Memoir of M. de Lamarck. *Zoological Philosophy: An Exposition with Regard to the Natural History of Animals* (ed. by Lamarck, J.B. Trans. H. Elliot), pp. 434–453. The University of Chicago Press, Chicago [from Cuvier, G. 1832. Éloge de M. De Lamarck, i — xxxj. lu à l'Académie des Sciences le 26 novembre 1832].

Darwin, C. (1859). *On the origin of species by means of natural selection, or, the preservation of favoured races in the struggle for life*. John Murray, London.

De Carvalho, M.R, Maisey, J.G, & Grande, L. (2004). Freshwater stingrays of the Green River Formation of Wyoming (Early Eocene), with the description of a new genus and species and an analysis of its phylogenetic relationships (Chondrichthyes: Myliobatiformes). *Bulletin of the American Museum of Natural History*, 284, 1–136.

Desmond, A. (1982). *Archetypes and ancestors: Palaeontology in Victorian London 1850–1875*. Frederick Muller, London.

Desmond, A. (1994). *Huxley: the devil's disciple*. Michael Joseph, London.

De Pinna, M.C.C. (1991). Concepts and tests of homology in the cladistic paradigm. *Cladistics*, 7, 367–394.

De Pinna, M.C.C. (1999). Species concepts and phylogenetics. *Reviews in Fish Biology and Fisheries*, 9, 353–373.

De Queiroz, A. (2005). The resurrection of oceanic dispersal in historical biogeography. *Trends in Ecology and Evolution*, 20, 68–73.

Di Gregorio, M.A. (1995). A wolf in sheep's clothing: Carl Gegenbaur, Ernst Haeckel, the vertebral theory of the skull, and the survival of Richard Owen. *Journal of the History of Biology*, 28, 247–280.

Di Gregorio, M.A. (2002). Refections of a non-political nautralist: Ernst Haeckel, Wilhelm Bleek, Freidrich Müller and the meaning of language. *Journal of the History of Biology*, 35, 79–109.

Dobbs, D. (2005). *Reef madness: Charles Darwin, Alexander Agassiz, and the meaning of coral*. Pantheon, New York.

Donoghue, M.J. & Moore, B.R. (2003). Toward an integrative historical biogeography. *Intergrated Comparative Biology*, 43, 261–270.

Donoghue, P.C.J. & Smith, M.P. (2003). *Telling the evolutionary time: Molecular clocks and the fossil record*. Taylor & Francis, London, New York.

Dunn, E.R. (1922). A suggestion to zoogeographers. *Science*, 56, 336–338.

Ebach, M.C. (1999). Paralogy and the centre of origin concept. *Cladistics*, 15, 387–391.

Excoffer, L. (2004). Analytical methods in phylogeography and genetic structure. *Molecular Ecology*, 13, 729–980.

Fernholm, B., Bremer, K.G., Jornvall, H. (1989). *The hierarchy of life: molecules and morphology in phylogenetic analysis*. Proceedings from Nobel symposium 70 held at Alfred Nobel's Bjorkborn, Karlskoga, Sweden, August 29–September 2, 1988. Excerpta Medica, Amsterdam & Oxford.

Finlayson, C. (2005). Biogeography and evolution of the genus *Homo*. *Trends in Ecology & Evolution*, 20, 457–463.

Franz, V. (Ed.) (1943–1944). *Ernst Haeckel: Sein Leben, Denken und Wirken* 2 vols., Wilhelm Gronau, Jena.

Gadow, H. (1913). *The Wanderings of Animals.* Cambridge, Cambridge University Press, Cambridge.

Gasman, D. (1971 [2004]). *The Scientific Origins of National Socialism.* Transaction Publishers, New Brunswick.

Gasman, D. (1998). *Haeckel's Monism and the Birth of Fascist Ideology.* Studies in Modern European History, 33, Peter Lang Publishing Inc, New York.

Ghiselin, M.T. (1992). Review of: Winsor, M.P. 1991. *Reading the shape of nature: Comparative zoology at the Agassiz Museum.* Chicago & London: University of Chicago Press. *Quarterly Review of Biology,* 67, 505–506.

Gladfelter, E.H. (2002). *Agassiz's Legacy: Scientists' Refl ctions on the Value of Field Experience.* Oxford University Press, Oxford, New York.

Goldschmidt, R.B. (1956). *Portraits from memory: Recollections of a zoologist.* University of Washington Press, Seattle, Washington.

Gould, S.J. (1973). Systematic pluralism and the uses of history. *Systematic Zoology,* 22, 322–324.

Gould, S.J. (1977). *Ontogeny and phylogeny.* Belknap Press of Harvard University Press, Cambridge, Massachusetts.

Gould, S.J. (1979). Agassiz' later, private thoughts on evolution: His marginalia in Haeckel's *Natü liche Schöpfungs -geschichte* (1868). *Two Hundred Years of Geology in America: Proceedings of the New Hampshire Bicentennial Conference on the History of Geology* (ed. by Schannder, C.J.), pp. 277–282. Hanover University Press of New England, Hanover, New Hampshire.

Gould, S.J. (2003). *Abscheulich!* (Atrocious). *I Have Landed,* pp. 305–320. Vintage Books.

Gray, A. & Adams, C.B. (1852). *Elements of Geology.* Harper & Brothers, New York.

Grehan, J.R. (1988). Biogeographic homology. Ratites and the southern beeches. *Revista di Biologia-Biology Forum,* 81, 577–587.

Grehan, J.R. (1991). A panbiogeographic perspective for pre-Cretaceous Angiosperm-Lepidoptera coevolution. *Australian Systematic Botany,* 4, 91–110.

Grehan, J.R. (2001). Panbiogeograf a y biogeograf a de la vida. Introducion a la biogeograf a: Teorias, conceptos, metodos y aplicaciones (ed. by Llorente, J. & J.J. Morrone), pp. 181–196. Universidad Nacional Autónoma de México, Mexico City.

Grehan, J.R. & Ainsworth, R. (1985). Orthogenesis and evolution. *Systematic Zoology,* 34, 174–192.

Haeckel, E. (1860). Abbildung und Diagnosen neuer Gattungen und Arten von lebenden Radiolarien des Mittelmeers. *Monatsbericht der Köi glichen Akademie der Wissenschaften Berlin,* 1860, 835–845.

Haeckel, E. (1861). *De Rhizopodum finibus et ordinibus: dissertatio, ...* Berolini.

Haeckel, E. (1862). *Die Radiolarien: (Rhizopoda Radiaria): Eine Monographie.* Berlin. i–xiv, 1–572 p, Atlas, 35 pls.

Haeckel, E. (1863). Über die Entwicklungstheorie Darwins. *Öffentlicher Vortrag in der Allgfemeinen Versammlung deutscher Naturforscher und Ärzte zu Steettin, am 19. 9. 1862 (Amtlicher Bericht üe r die 37. Versammlung S. 17).*

Haeckel, E. (1866). *Generelle Morphologie der Organismen: Allgemeine Grundzü e der organischen Formen-Wissenschaft, mechanisch begrüde t durch die von C. Darwin reformirte Decendenz-Theorie.* Berlin. 2 volumes.

Haeckel, E. (1868). *Natü liche Schöfun gsgeschichte. Gemeinverstädl iche wissenschaftliche Vortrü e üe r die Entwicklungslehre im Allgemeinen und diejenige von Darwin, Goethe und Lamarck im Besonderen üe r die Anwendung derselben auf den Ursprung des Menschen und andere damit zusammenhäg ende Grundfragen der Naturwissenschaft.* Berlin.

Haeckel, E. (1870). *Natürliche Schöpfungsgeschichte (2nd ed.). Gemeinverständliche wissenschaftliche Vorträge über die Entwicklungslehre im Allgemeinen und diejenige von Darwin, Goethe und Lamarck im Besonderen über die Anwendung derselben auf den Ursprung des Menschen und andere damit zusammenhängende Grundfragen der Naturwissenschaft.* Berlin.

Haeckel, E. (1872). *Die Kalkschwä me.* Berlin.

Haeckel, E. (1874a). *Natü liche Schöfungs geschichte (5th ed.). Gemeinverstädl iche wissenschaftliche Vortrя e ße r die Entwicklungslehre im Allgemeinen und diejenige von Darwin, Goethe und Lamarck im Besonderen.* Berlin.

Haeckel, E. (1874b). *Anthropogenie oder Entwickelungsgeschichte des Menschen: Gemeinverstandliche wissenschaftliche Vortrage uber die Grundzuge der menschlichen Keimes- und Stammes-genchichte (3rd ed.).* Wilhelm Engelmann, Leipzig.

Haeckel, E. (1876a). *The History of Creation, or, the Development of the Earth and Its Inhabitants by the Action of Natural Causes: Doctrine of Evolution in General, and of That of Darwin, Goethe, and Lamarck in Particular* (translation revised by E. Ray Lankester). Henry S. King, London,

Haeckel, E. (1876b). Ziele und Wege der Heutigen Entwickelungsgeschichte. *Jenaische Zeitschrift fü Medizin und Naturwissenschaften* 10, suppl., 99 pp.

Haeckel, E. (1879a). *Natü liche Schöf ungsgeschichte. Gemeinverstädl iche wissenschaftliche Vortrя e ße r die Entwicklungslehre im Allgemeinen und diejenige von Darwin, Goethe und Lamarck im Besonderen.* Berlin.

Haeckel, E. (1879b). *The Evolution of Man: a Popular Exposition of the Principal Points of Human Ontogeny and Phylogeny (Anthropogenie oder Entwicklungsgeschichte des Menschen)*, Kegan Paul, London. (Translated from the German: *Anthropogenie oder Entwicklungsgeschichte des Menschen: Gemeinverstandliche wissenschaftliche Vortrage uber die Grundzuge der Menschlichen Keimes- und Stammes-geschichte.* Wilhelm Engelmann, Leipzig).

Haeckel, E. (1883). *The Evolution of Man: a Popular Exposition of the Principal Points of Human Ontogeny and Phylogeny (Anthropogenie oder Entwicklungsgeschichte des Menschen)*, D. Appleton, New York (Translated from the German: *Anthropogenie oder Entwicklungsgeschichte des Menschen: gemeinverstandliche wissenschaftliche Vortrage uber die Grundzuge der menschlichen Keimes- und Stammes-genchichte.* Wilhelm Engelmann, Leipzig).

Haeckel, E. (1887a). Report on the Radiolaria collected by H.M.S. Challenger during … 1873–76. *Report on the scientific results of the voyage of H.M.S. Challenger during the years 1872–76.* John Murray, Volume 18, part 40.

Haeckel, E. (1887b). *Die Radiolarien: (Rhizopoda Radiaria.) Eine Monographie.* 2 Theil, *Grundriss einer allgemeinen Naturgeschichte der Radiolarien,* & c. Berlin.

Haeckel, E. (1889). *Natü liche Schöfungs geschichte (8nd ed.). Gemeinverstädl iche wissenschaftliche Vortrя e ße r die Entwicklungslehre im Allgemeinen und diejenige von Darwin, Goethe und Lamarck im Besonderen (8th ed.).* Berlin.

Haeckel, E. (1891). *Anthropogenie oder Entwickelungsgeschichte des Menschen: Keimes- und Stammesgeschichte.* 4th ed., Wilhelm Engelmann, Leipzig.

Haeckel, E. (1893). Zur Phylogenie der australischen Fauna. Systematische Einleitung. In *Zoologische Forschungsreisen in Australien und dem Malayischen Archipel* (1893–1913), (ed. by Semon, R.), 1: i–xxiv.

Haeckel, E. (1894–1896). *Systematische Phylogenie: Entwurf eines natü lichen Systems der Organismen auf Grund ihrer Stammesgeschichte.* Berlin, 3 v: Vol. 1 [1894] *Systematische Phylogenie der Protisten und Pflnz en;* Vol. 2 [1895] *Systematische Phylogenie der Wirbellosen Thiere (Invertebrata);* Vol. 3 [1896] *Systematische Phylogenie der Wirbelthiere (Vertebrata).* Berlin.

Haeckel, E. (1902). *Gemeinverstüdlic he Vorträ e und Abhandlungen aus dem Gebiete der Entwickelungslehre.* Bonn.

Haeckel, E. (1904). *The Wonders of Life: A Popular Study of Biological Philosophy.* Watts, London. (Supplementary volume to *The Riddle of the Universe.* Translated from: *Die Lebenswunder,* 1904).

Haeckel, E. (1906). *Prinzipien der generellen Morphologie der Organismen: Wörtlicher Abdruck eines Teiles der 1866 erschienen Generellen Morphologie (Allgemeine grundzüge der organischen formen-wissenschaft mechanisch begründet durch die von Charles Darwin reformierte deszendenz-theorie)* ... Mit dem porträt des verfassers. G. Reimer, Berlin.

Haeckel, E. (1907). *Das Menschenproblem und die Herrentiere von Linné.* Frankfurt, Frankfurter Verlag.

Haeckel, E. (1909). *The Riddle of the Universe at the Close of the Nineteenth Century* (translated by J. McCabe, &c), 5th ed., Watts & Co., London, xvi, 142 pp.

Haeckel, E. (1925). *The History of Creation: Or, the Development of the Earth* ... . Henry S. King, London.

Haffer, J. (2003). Wilhelm Meise (1901–2002), ein führender Ornithologe Deutschlands im. 20. Jahrhundert. *Verhandlungen des Vereins fü naturwissenschaftliche Heimatforschung zu Hamburg,* 40, 117–140.

Haffer, J., Rutschke, E. & Wunderlich, K. (Eds.) (2000). Erwin Stresmann (1889–1972) — Leben und Werk eines Pioniers der wissenschaftlichen Ornithologie. *Acta Historica Leopoldiana,* 34, 1–465.

Haller, J.S. (1970). The species problem: nineteenth-century concepts of race inferiority in the origin of man controversy. *American Anthropologist,* 72, 1319–1329.

Heads, M.J. (1985). On the nature of ancestors. *Systematic Zoology,* 34, 205–215.

Heads, M.J. (2005a). Dating nodes on molecular phylogenies: a critique of molecular biogeography. *Cladistics,* 21, 62–78.

Heads, M.J. (2005b). Towards a panbiogeography of the seas. *Biological Journal of the Linnean Society,* 84, 675–723.

Heads, M.J. (2005c). The history and philosophy of panbiogeography. In *Regionalizació Biogeográfica en Iberoamérica y Tpic os Afines* (ed. by Llorente, J. & Morrone, J.J), pp. 67–123. Universidad Nacional Autónoma de México, México.

Heberer, G. (1968). *Der gerechtfertigte Haeckel: Einblicke in seine Schriften aus Anlass des Erscheinens seines Hauptwerkes G enerelle Morphologie der Organismen" vor 100 Jahren.* Gustav Fischer Verlag, Stuttgart.

Heilborn, A. (1920). *Die Lear-Tragdie Ernst Haeckels.* Hamburg, Berlin.

Hehn, V. (1885). *The Wanderings of Plants and Animals from Their First Home.* London.

Hennig, W. (1950). *Grundzü e einer Theorie der phylogenetischen Systematik.* Deutsche Zentralverlag, Berlin.

Hennig, W. (1966). *Phylogenetic Systematics.* University of Illinois Press, Urbana.

Hertler, C. & Weingarten, M. (2001). Ernst Haeckel (1834–1919). In *Darwin & Co.: Eine Geschichte der Biologie in Portraits* I (ed. by Jahn, I. & Schmitt, M.), pp. 434–455, 535–537. C.H. Beck, München.

Hofsten, N. V. (1916). Zur älteren Gesschichte des Diskontinuitätsproblems in der Biogeographie. *Zoologische Annalen Zeitschrift fü Geschichte der Zoologie,* 7, 197–353.

Holmes, S.J. (1944). Recapitulation and its supposed causes. *Quarterly Review of Biology,* 19, 319–331.

Horton, D. (1973). The concept of zoogeographic subregions. *Systematic Zoology,* 22, 191–195.

Hoßfeld, U. (2004). The travels of Jena zoologist in the Indo-Mayalan region. *Proceedings of the Californian Academy of Sciences,* 55, 77–105.

Hoßfeld, U. & Olsson, L. (2003). The road from Haeckel: the Jena tradition in evolutionary morphology and the origins of "evo-devo." *Biology and Philosophy,* 18, 285–307.

Hubbs, C.L. (1945). Reviews and comments: *Foundations of Plant Geography.* by Stanley A. Cain. *American Naturalist,* 79, 176–178.

Hull, D.L. 1973. *Darwin & His Critics: the Reception of Darwin's Theory of Evolution by the Scientific Community.* Harvard University Press, Cambridge, Massachusetts.

Humphries, C.J. & Parenti, L.R. (1999). *Cladistic Biogeography.: Interpreting Patterns of Plant and Animal Distributions* 2nd ed.. Oxford University Press, Oxford.

Huxley, C.R., Lock, J.M. & Cutler, D.F. (1998). *Chorology, Taxonomy and Ecology of the Floras of Africa and Madagascar — from the Frank White Memorial Symposium held in the Plant Sciences Department, Oxford University on Sept 26–27th 1996 by the Linnean Society of London, the Royal Botanic Gardens, Kew and Wolfson College, Oxford.* The Royal Botanic Gardens, Kew.

Huxley, T.H. (1868). On the classif cation and distribution of the Alectoromorphae and Heteromorphae. *Proceedings of the Zoological Society,* (1868), 294–319 [reprinted in *Scientific Memoirs* 3, 346–373].

Huxley, T.H. (1869). [Review of] *The Natural History of Creation. The Academy,* 1, 12–14, 40–43.

Huxley, T.H. (1878). Evolution in Biology. *Encyclopaedia Britannica.* 9th ed., 8, 744–751 [reprinted in *Science and Culture, and Other Essays,* 1881 and *Collected Essays* 2 (*Darwinian*), 187–226, 1893].

Huxley, T.H. (1894). Owen's position in the history of anatomical science. In *The Life of Richard Owen* (ed. by Owen, R.), 2, 273–332 [reprinted in *Scientific Memoirs* 4, 658–689].

Hyatt, A. (1894). Phylogeny of an acquired character. *Proceedings of the American Philosophical Society, Philadelphia,* 32, 349–647.

Jacobi, A. (1900). Lage und Form biogeographischer Gebiete. *Zeitschrift der Gesellschaft fü Erdkunde zu Berlin,* 35, 47–238.

Janvier, P. (1996 [1998]). *Early Vertebrates.* Clarendon Press, Oxford.

Janvier, P. (2003). Armoured f sh from deep time: From Hugh Miller's insights to current questions of early vertebrate evolution. *Celebrating the Life and Times of Hugh Miller. Scotland in the Early 19th Century* (ed. by Borley, L.), pp. 177–196. Cromarty Arts Trust, Cromarty.

Junker T. (1991). Heinrich Georg Bronn und die Entstehung der Arten. *Sudhoffs Archiv fü Geschichte der Medizin und der Naturwissenschaften,* 75, 180–208.

Junker, T. (2003). Ornithology and the genesis of the synthetic theory of evolution. *Avian Science,* 3, 65–73.

Just, T. (1947). Geology and plant distribution. *Ecological Monographs,* 17, 127–137.

Keller, C. (1929). Louis Agassiz und seine Stellung in der Biologie. *Verhandlungen der Naturforschenden Gesellschaft in Basel,* 40, 43–52.

Kinch, M.P. (1980). Geographical distribution and the origins of life: the development of early nineteenth century British explanations. *Journal of the History of Biology,* 13, 91–119.

Kiriakoff, S. (1954). Chorologie et systématique phylogénétique. *Bulletin et Annales de la Société Royale Belge d'Entomologie,* 90, 185–198.

Kirchengast, S. (1998). Ernst Haeckel und seine Bedeutung für die Entwicklung der Paläanthropologie. In Ernst Haeckel — Ein Plädoyer für die wirbellosen Tiere und die biologische Systemtik. *Weltrts el und Lebenswunder. Ernst Haeckel — Wirk, Wirkung und Folgen.* Stapf a 56, N.F. 131, 169–184.

Klemm, P. (1969). *Ernst Haeckel. Der Ketzer von Jena.* Urania, Leipzig, Jena, Berlin.

Koerner, K. (1987). On Schleicher and trees. In *Biological Metaphor and Cladistic Classification* (ed. by Hoenigswald, H.M. & Wiener, L.F.), pp. 109–113. University of Pennsylvania Press, Philadelphia.

Koerner, K. (1989). Schleichers Einfuss auf Haeckel: Schlaglichter auf die Abhängigkiet zwischen linguistischen und biologischen Therion im 19. Jahrundert. In *Practising Linguistic Hisotriography: Selected Essays* (ed. by Koerner, K.), pp. 211–214. John Benjamins, Philadelphia.

Kohlbrugge, J.H.F. (1911). Das biogenetische Grundgesetz. Eine historische Studie. *Zoologischer Anzeiger,* 37, 447–453.

Krumbach, T. (1919). Die schriften Ernst Haeckel. *Die Naturwissenschaft,* 7, 961–966.

Laurent, G. (1987). *Paléontologie et évolution en France de 1800 a 1860: Une histoire des idees de Cuvier et Lamarck a Darwin.* Éditions du CTHS, Paris.

Laurent, G. (1997). Paléontologie et évolution: Etat de la question en 1850 d'après l'œuvre de Heinrich-Georg Bronn (1800–1862). *De la géologie à son histoire* (ed. by Gohau, G. & Gaudant, J.), pp. 175–188. Comité des travaux historiques et scientifques, Mémoires de la section des sciences 13, Paris.

Laurent, G. (2001). *La naissance du transformisme: Lamarck entre Linne et Darwin.* Vuibert-Adapt, Paris.

Lebedkin, S. (1936). The recapitulation problem. Part. I. *Bulletin de l' Institut Lesshaft,* pp. 391–417.

Lebedkin, S. (1937). The recapitulation problem. Part. II. *Bulletin de l' Institut Lesshaft,* pp. 561–594.

Le Conte, J. (1888). *Evolution in its Relation to Religious Thought.* D. Appleton and Company, New York.

Le Conte, J. (1903). *The Autobiography of Joseph LeConte.* D. Appleton and Company, New York.

Le Conte, J. (1905). *Evolution in its Relation to Religious Thought.* D. Appleton and Company, New York.

Lomolino, M.V., Sax, D.F., & Brown, J.H. (2004). *Foundations of Biogeography: a Collection of Seminal Papers.* University of Chicago Press, Chicago.

Llorente, J., Morrone, J.J., Bueno, A., Pérez, R. Vitoria, Á., & Espinosa, D. (2000). Historia del desarrollo y la recepción de las ideas panbiogeográfcas de Léon Croizat. *Academia Colombiana de Ciencias Exactas, Fisicas y Naturales,* 24(93), 549–577.

Lovejoy, A.O. (1909a). The argument for organic evolution before *The Origin of Species.* I. *The Popular Science Monthly,* 75, 499–514.

Lovejoy, A.O. (1909b). The argument for organic evolution before *The Origin of Species.* II. *The Popular Science Monthly,* 75, 537–549.

Lovejoy, N.R. (1996). Systematics of myliobatoid elasmobranchs: with emphasis on the phylogeny and historical biogeography of neotropical freshwater stingrays (Potamotrygonidae: Rajiformes). *Zoological Journal of the Linnean Society,* 117, 207–257.

Lovejoy, N.R. (1997). Stingrays, parasites, and Neotropical biogeography: a closer look at Brooks et al.'s hypothesis concerning the origins of Neotropical freshwater rays. *Systematic Biology,* 46, 218–230.

Løvtrup, S. (1987). On species and other taxa. *Cladistics,* 3, 157–177.

Lurie, E. (1959). Louis Agassiz and the races of Man. *Isis,* 45, 227–242.

Lurie, E. (1960). *Louis Agassiz: A Life in Science.* University of Chicago Press, Chicago, Ill. [reprint 1988].

Lyell, C. (1830–1832). *Principles of Geology, Being an Attempt to Explain the Former Changes of the Earth's Surface.* London, Volumes 1–3 [Reprinted by University of Chicago Press, 1990–1991].

Lynch, J.M. (2003). Introduction. *Agassiz on Evolution,* pp. v–xxi, Thoemmes Press, Bristol.

Marcou, J. (1896 [1972]). *Life, Letters, and Works of Louis Agassiz, &c.* Gregg International Publishers, Westmead, Hants. (Facsimile reprint of the edition published by Macmillan, London and New York, 1896, 2 Vol. in 1).

Matthew, W.D. (1915). Climate and evolution. *Annals of the New York Academy of Sciences,* 24, 171–318.

Mayr, E. (1946). History of the North American bird fauna. *Wilson Bulletin,* 58, 3–41 [reprinted and modif ed in Mayr 1976 [1997]. *Evolution and the Diversity of Life,* The Belknap Press of Harvard University, pp. 565–588].

Mayr, E. (1965). What is a fauna? *Zoologisches Jahrbuch, Abteilung Systematik,* 92, 473–486 [reprinted and slightly modif ed in Mayr, 1976 [1997]. *Evolution and the Diversity of Life,* Belknap Press of Harvard University, pp. 552–564].

Mayr, E. (1982a). Review of Nelson, G., & D.E. Rosen 1981 (Eds.). *Vicariance Biogeography: a Critique.* Columbia University Press, New York. *The Auk,* 99, 618–620.

Mayr, E. (1982b). *The Growth of Biological Thought: Diversity, Evolution, and Inheritance.* Harvard University Press, Cambridge, MA.

Mayr, E. (1994). Recapitulation reinterpreted: The somatic program. *Quarterly Journal of Biology,* 69, 223–232.

Mayr, E. (1997). Reminiscences of Erwin Stresemann: teacher and friend. In *W e Must Lead the Way on New Paths," The Work of Hartert, Stresemann, Ernst Mayr — International Ornithologists,* (ed. by Haffer, J.), Ornithologen-Briefe des 20. Jahrhunderts. *Ökologie der Vğ el,* 19, 848–855.

Mayr, E. (1999). Thoughts on the evolutionary synthesis in Germany. In *Die Entstehung der Synthetischen Theorie: Beiträ e zur Geschichte der Evolutionsbiologie in Deutschland 1930–1950* (ed. by Junker, T. & Engels, E.-M.), pp. 19–30. Verlag für Wissenschaft und Bildung, Berlin.

Mayr, E. & Bock, W.J. (2002). Classif cations and other ordering systems. *Journal of Zoological Systematics and Evolutionary Research,* 40, 169–194.

Meise, W. & Hennig, W. (1932). Die Schlangengattung *Dendrophis. Zoologischer Anzeiger,* 99, 273–297.

Meise, W. & Hennig, W. (1935). Zur Kenntnis von *Dendrophis* und *Chrysopelea. Zoologischer Anzeiger,* 109, 138–150.

Merriam, C.H. & Stejneger, L.H. (1890). *Results of a Biological Survey of the San Francisco Mountain Region and Desert of the Little Colorado, Arizona.* North American Fauna, Washington, D.C., No. 3. 136 pp.

Meyer, A.W. (1935). Some historical aspects of the recapitulation idea. *Quarterly Review of Biology,* 10, 379–396.

Meyer, A.W. (1936). Haeckel, Lamarck and "Intellectual Inertia." *American Naturalist,* 70, 494–497.

Moore, J. & Desmond, A. (2004). Introduction. In C. Darwin, *The Descent of Man: Selection in Relation to Sex.* Penguin Books, London.

Morris, P.J. (1997). Louis Agassiz's arguments against Darwinism in his additions to the French translation of the Essay on Classif cation. *Journal of the History of Biology,* 30, 121–134.

Morrone, J.J. (2000). El tiempo de Darwin y el espacio de Croizat: Rupturas epistémicas en los estudios evolutivos. *Ciencia,* 51, 39–46.

Morrone, J.J. (2001). Homology, biogeography and areas of endemism. *Diversity and Distributions,* 7, 297–300.

Morrone, J.J. (2002). Biogeographic regions under track and cladistic scrutiny. *Journal of Biogeography,* 29, 149–152.

Morrone, J.J. (2004). *Homologá biogeogrfic a: Las coordenadas espaciales de la vida.* Cuadernos del Instituto de Biología 37, Instituto de Biología, UNAM, México.

Morrone, J.J. (2005a). Cladistic biogeography: identity and place. *Journal of Biogeography,* 32, 1281–1286.

Morrone, J.J. (2005b). Erratum. *Journal of Biogeography,* 32, 1505.

Müller, I. (1998). Historische Grundlagen des Biogenetischen Grundgesetzes. Ernst Haeckel — Ein Plädoyer für die wirbellosen Tiere und die biologische Systemtik. *Welträtsel und Lebenswunder. Ernst Haeckel — Wirk, Wirkung und Folgen. Stapfia* 56, N.F. 131, 119–130.

Nelson, G. (1969). Gill arches and phylogeny of f shes, with notes on the classif cation of vertebrates. *Bulletin of the American Museum of Natural History,* 141, 475–552.

Nelson, G. (1973). Comments on Leon Croizat's biogeography. *Systematic Zoology,* 22, 312–320.

Nelson, G. (1978a). Ontogeny, phylogeny, paleontology, and the biogenetic law. *Systematic Zoology* 27: 324–345.

Nelson, G. (1978b). From Candolle to Croizat: comments on the history of biogeography. *Journal of the History of Biology,* 11, 269–305.

Nelson, G. (1979). Cladistic analysis and synthesis: Principles and def nitions, with a historical note on Adanson's *Familles des Plantes. Systematic Zoology,* 28, 1–21.

Nelson, G. (1983). Vicariance and cladistics: Historical perspectives with implications for the future. In *Evolution, Time and Space: The Emergence of the Biosphere* (ed. by Sims, R.W., Price J.H. & Whalley, P.E.S.), pp. 469–492. Academic Press, London.

Nelson, G. (1985). A decade of challenge: the future of biogeography. In *Plate Tectonics and Biogeography. Earth Sciences History* (ed. by Leviton, A.E. & Aldrich, M.L.), *Journal of the History of the Earth Sciences Society,* 4, 187–196.

Nelson, G. (1989). Species and taxa: Systematics and evolution. *Speciation and Its Consequences* (ed. by Otte, D. & Endler, J.), pp. 60–81. Sinauer, Sunderland.

Nelson, G. (1994). Homology and systematics. *Homology: the Hierarchical Basis of Comparative Biology* (ed. by Hall, B.K.), pp. 101–149. Academic Press, San Diego.

Nelson, G. & Ladiges, P.Y. (2001). Gondwana, vicariance biogeography, and the New York School revisited. *Australian Journal of Botany,* 49, 389–409.

Nelson, G. & Platnick, N.I. (1981). *Systematics and Biogeography: Cladistics and Vicariance.* Columbia University Press, New York.

Nelson, G. & Platnick, N.I. (1984). Systematics and evolution. In *Beyond Neodarwinism: an Introduction to the New Evolutionary Paradigm,* (ed. by Ho, M.-W. & Saunders, P.T.), pp. 143–158. Academic Press, London.

Nelson, G. & Rosen, D.E. (1981). *Vicariance Biogeography: a Critique.* Columbia University Press, New York.

Nordenskiöld, E. (1936). *The History of Biology: A Survey.* Translated from the Swedish by Leonard Bucknall Eyre. New ed. Tudor, New York.

Nyhart, L.K. (1995). *Biology Takes Form: Animal Morphology and the German Universities, 1800–1900.* University of Chicago Press, Chicago.

Nyhart, L.K. (2002). Learning from history: morphology's challenge in Germany ca. 1900. *Journal of Morphology,* 252, 2–14.

Oppenheimer, J.M. (1959). Embryology and evolution: Nineteenth century hopes and twentieth century realities. *Quarterly Journal of Biology,* 34, 271–277.

Oppenheimer, J.M. (1987). Haeckel's variations on Darwin. In *Biological Metaphor and Cladistic Classification* (ed. by Hoenigswald, H.M. & Wiener, L.F.), pp. 123–135. University of Pennsylvania Press, Philadelphia.

Ortmann, A.E. (1902a). The geographical distribution of freshwater Decapods and its bearing upon ancient geography. *Proceedings of the American Philosophical Society,* 41, 267–400.

Ortmann, A.E. (1902b). Biogeographical regions. *American Naturalist,* 36, 157–159.

Page, R.D.M. (1989). New Zealand and the new biogeography. *New Zealand Journal of Zoology,* 16, 471–493.

Papavero, N., Llorente-Bousquets, J. & Abe, J.M. (1997). *Fundamentos de biología comparada: (a trá es de la teoria intuitiva de conjuntos)* Vol.1 De Platón a Haeckel. Universidad Nacional Autonoma de Mexico, Mexico.

Parenti, L. (1991). Ocean basins and the biogeography of freshwater f shes. *Australian Systematic Botany,* 4, 137–149.

Patterson, C. (1977). The contribution of paleontology to teleostean phylogeny. *Major Patterns in Vertebrate Evolution* (ed. by Hecht, M.K., Goody, P.C. & Hecht, B.M.), pp. 579–643. Plenum Press, New York.

Patterson, C. (1980). Cladistics. *Biologist,* 27, 234–240.

Patterson, C. (1981a). Agassiz, Darwin, Huxley, and the fossil record of teleost f shes. *Bulletin of the British Museum (Natural History), Geology,* 35, 213–224.

Patterson, C. (1981b). Methods of paleobiogeography. In *Vicariance Biogeography: A Critique* (ed. by Nelson, G. & Rosen, D.E.), pp. 446–500, Columbia University Press, New York.

Patterson, C. (1983). Aims and methods in biogeography. In *Evolution, Time and Space: The Emergence of the Biosphere* (ed. by Sims, R.W., Price, J.S. & Whalley, P.E.S.), pp. 1–28. Academic Press, London.

Patterson, C. (1989). Phylogenetic relations of major groups: Conclusions and prospects. In *The Hierarchy of Life* (ed. by Fernholm, B., Bremer, K. & Jôrnvaîl, H.), Nobel Symposium 70, pp. 471–488. Excerpta Medica, Amsterdam.

Platnick, N.I. & G. Nelson. (1989). Spaning-tree biogeography: shortcut, detour, or dead-end? *Systematic Zoology,* 37, 410–419.

Prichard, J.C. (1813). *Researches into the Physical History of Mankind.* John and Arthur Arch, London.

Prichard, J.C. (1826). *Researches into the Physical History of* Mankind. John and Arthur Arch, London, 2nd Edition, 2 vols.

Prichard, J.C. (1843). *The Natural History of Man.* John and Arthur Arch, London.

Radl, E. (1930). *The History of Biological Theories.* Translated and adapted by E.J. Hatf eld, &c. London.

Richards, R. (1987). *Darwin and the Emergence of Evolutionary Theories of Mind and Behaviour.* University of Chicago Press, Chicago, London.

Richards, R. (1992). *The Meaning of Evolution: the Morphological Construction and Ideological Reconstruction of Darwin's Theory.* University of Chicago Press, Chicago, London.

Richards, R. (2002). The linguistic creation of Man: Charles Darwin, August Schleicher, Ernst Haeckel, and the missing link in 19th-century evolutionary theory. In *Experimenting in Tongues: Studies in Science and Language,* (ed. by Dörres, M.), Stanford University Press, Stanford.

Richards, R. (2004). If this be heresy: Haeckel's conversion to Darwinism. In *Darwinian Heresies,* (ed. by Lusting, A., Richards, R.J. & Ruse, M.), pp. 101–130. Cambridge University Press, Cambridge.

Richards, R. (2005). The aesthetic and morphological foundations of Ernst Haeckel's evolutionary project. *The Many Faces of Evolution in Europe, 1860–1914,* (ed. by M. Kemperink and P. Dassen), pp. 1–16. Peeters, Amsterdam.

Richards, R. (2006). *The Tragic Sense of Life: Ernst Haeckel and the Battle over Evolution in Germany.* University of Chicago Press, Chicago, London.

Richardson, R.A. (1981). Biogeography and the genesis of Darwin's ideas on transmutation. *Journal of the History of Biology,* 14, 1–41.

Richardson, M.K. & Keuck, G. (2002). Haeckel's ABC of evolution and development. *Biological Reviews,* 77, 495–528.

Rieppel, O. (1988). Louis Agassiz (1807–1873) and the reality of natural groups. *Biology and Philosophy*, 3, 29–47.

Roberts, J.H. (1982). *Darwinism and the Divine in America. Protestant Intellectuals and Organic Evolution, 1859–1900.* The University of Wisconsin Press, Madison, London.

Roger, J. (1983). Darwin, Haeckel et les Français. *De Darwin au Darwinisme: science et idéologie* (ed. by Conry Y.), pp. 149–165. J. Vrin, Paris.

Rudwick (1972 [1985]). *The Meaning of Fossils: Episodes in the History of Palaeontology.* Macdonald, London.

Rupp-Eisenreich, B. (1996). Haeckel, Ernst Hienrich 1834–1919. In *Dictionnaire de Darwinisme et de l'evolution* (ed. by Tort, P.), Volume 2, pp. 2073–2114. Presses Universitaires de France, Paris.

Ruse, M. (1997). *Monad to Man: the Concept of Progress in Evolutionary Biology.* Harvard University Press, Cambridge, Massachusetts.

Russell, E.S. (1916). *Form and Function: a Contribution to the History of Animal Morphology.* University of Chicago Press, Chicago.

Schleicher, A. (1850). *Die Sprachen Europas in Systematischer Uebersicht,* H.B. Konig, Bonn.

Schleicher, A. (1853). Die ersten Spaltungen des indogermanischen Urvolkes. *Allgemeine Monatsschrift fü Wissenschaft und Literatur,* 1853, 876–787.

Schleicher, A. (1860). *Die Deutsche Sprache.* Stuttgart.

Schleicher, A. (1863). *Die Darwinische Theorie und die Sprachwissenschaft. Offenes Sendschreiben an Herrn Dr. Ernst Haeckel.* Weimar.

Schleicher, A. (1865). *Über die Bedeutung der Sprache fü die Naturgeschichte des Menschen.* Böhlau, Weimar.

Schleicher, A. (1869). *Darwinism tested by the Science of Language.* London. [Reprinted in Koerner, K. (ed.) 1983. *Linguistics and Evolutionary Theory: Three Essays by August Schleicher, Ernst Haeckel, and Wilhelm Bleek.* John Benjamins, Philadelphia.]

Schmidt, K.P. (1954). Faunal realms, regions, and provinces. *Quarterly Journal of Biology,* 29, 322–331.

Schmitt, M. (2001). Willi Hennig (1913–1976). *Darwin & Co.: Eine Geschichte der Biologie in Portraits* II (ed. by Jahn, I. & Schmitt, M.), pp. 316–343, 541–546. C.H. Beck, München.

Sclater, P.L. (1858). On the geographical distribution of the class Aves. *Journal of the Linnaean Society of London, Zoology,* 2, 130–145.

Seibold, I. & Seibold, E. (1997). Heinrich Georg Bronn; ein Brief von 1855 zur Evolutionstheorie vor Darwins Werk von 1859. *Geologische Rundschau,* 86, 518–521.

Shumway, W. (1932). The Recapitulation Theory. *Quarterly Review of Biology,* 7, 93–99.

Simpson, G.G. (1943). Criteria for genera, species, and subspecies in zoology and paleozoology. *Annals of the New York Academy of Sciences,* 44, 143–178.

Stauffer, R.C. (1957). Haeckel, Darwin, and Ecology. *Quarterly Journal of Biology,* 32, 138–144.

Stegmann, B. (1938). Principes généraux des subdivisions ornithogeographiques de la region paléarctique. *Faune de l'URSS,* n.s. 19, *Oiseaux,* 1(2).

Stephens, L.D. (2000). *Science, Race, and Religion in the American South: John Bachman and the Charleston Circle of Naturalists, 1815–1895.* University of North Carolina Press, Chapel Hill.

Stevens, P.F. (1994). *The Development of Biological Systematics: Antoine-Laurent de Jussieu, Nature and the Natural System.* Columbia University Press, New York.

Stocking, G.W. (1973). Introduction. In James Cowles Prichard. In *Researches into the Physical History of Mankind,* 1813. University of Chicago Press, Chicago.

Stresemann, E. (1939). Die Vögel von Celebes. Zoogeographie. *Journal fü Ornithologie,* 87, 312–425.

Suessenguth, K. (1938). *Neue Ziele der Botanik*. Munich.

Swainson, W. (1835). *A Treatise on the Geography and Classification of Animals*. Longman, Brown, Green, & Longmans, London.

Tiedemann, F. (1808). *Zoologie: zu seinen Vorlesungen entworfen*. Landshut, Heidelberg.

Tort, P. (1996). Haeckel — Liste Chronologique des Travaux Publié. *Dictionnaire de Darwinisme et de l'evolution* (ed. by Tort, P.), Volume 2, pp. 2114–2121. Presses Universitaires de France, Paris.

Traub, L. (1993). Evolutionary ideas and 'empirical' methods: The analogy between language and species in works by Lyell and Schleicher. *British Journal for the History of Science*, 26, 171–193.

Ulrich, W. (1967). Ernst Haeckel: "Generelle Morphologie", 1866. *Zoologische Beiträ e*, N.F. 13, 165–212.

Ulrich, W. (1968). Ernst Haeckel: "Generelle Morphologie", 1866. *Zoologische Beiträ e*, N.F. 14, 213–311.

Uschmann, G. (1959). *Geschichte der Zoologie und der Zoologischen Anstalten in Jena 1779–1919*. Jena.

Uschmann, G. (1967a). 100 "Jahre Generelle Morphologie". *Biologische Rundschau*, 5, 241–252.

Uschmann, G. (1967b). Zur Geschichte der Stammbaum-Darstellungen. In *Gesammelte Vorträ e üe r moderne Probleme der Abstammungslehre* (ed. by Gersch, M.), Band II, pp. 9–30. Friedrich-Schiller Universität, Jena.

Uschmann, G. (1972). Haeckel, Ernst Heinrich Philipp August. In *Dictionary of Scientific Biography* (ed. by Gillipsie, C.C.), Volume VI, pp. 6–11. Jean Hachette–Joseph Hyrth.

Uschmann, G. (1983). Ernst Haeckel. *Biographie in Briefen*. Urania, Leipzig.

Volkmann, E. von. (1943). Ernst Haeckel veranlasste die Einladung Bismarks. *Sein Leben, Denken und Wirken*. Haeckel (ed. by Franz, V.), 1, 80–89. W. Gronau, Jena.

Voous, K.H. (1963). The concept of faunal elements or faunal types. *Proceedings of the 13th Ornithological Congress, Ithaca 1962*, 2, 1104–1108.

Wallace, A.R. (1876). *The Geographical Distribution of Animals, with a Study of the Relations of Living and Extinct Faunas As Elucidating the Past Changes of the Earth's surface*. Macmillan and Co., London.

Walls, L.D. (2003). *Emerson's Life in Science: The Culture of Truth*. Cornell University Press, New York.

Weindling, P. (1989). Ernst Haeckel, Darwinismus and the secularisation of nature. In *History, Humanity, and Evolution: Essays for John C. Greene* (ed. by Moore, J.R.), pp. 311–327. Cambridge University Press, Cambridge.

Wells, R.S. (1987). The life and growth of language: metaphors in biology and linguistics. In *Biological Metaphor and Cladistic Classification* (ed. by Hoenigswald, H.M. & Wiener, L.F.), pp. 37–80. University of Pennsylvania Press, Philadelphia.

Williams, D.M. (2004). Homology and homologues, cladistics and phenetics: 150 years of progress. In *Milestones in Systematics* (ed. by Williams, D.M. & Forey, P.L.), pp. 191–224. CRC Press, Florida.

Williams, D.M. & Ebach, M.C. (2004). The reform of palaeontology and the rise of biogeography: 25 years after 'Ontogeny, Phylogeny, Paleontology and the Biogenetic law' (Nelson, 1978). *Journal of Biogeography*, 31, 1–27.

Wilson, H.V. (1941). The recapitulation theory or biogenetic law in embryology. *American Naturalist*, 75, 20–30.

Winsor, M.P. (1979). Louis Agassiz and the species question. *Studies in History of Biology*, 3, 89–117.

Winsor, M.P. (1991). *Reading the Shape of Nature: Comparative Zoology at the Agassiz Museum*. University of Chicago Press, Chicago & London.

Winsor, M.P. (2000). Agassiz's notions of a Museum. In *Cultures and Institutions of Natural History* (ed. by Ghiselin, M.R. & Leviton, A.E.), pp. 249–271. California Academy of Sciences, San Francisco.

Zimmermann, E.A.W. (1778–1783). *Geographische Geschichte des Menschen, und der allgemein verbreiteten vierfü sigen Thiere, nebst einer hieher gehö igen zoologischen Weltcharte*. Leipzig.

Zimmermann, W. (1931). Arbeitsweise der botanischen Phylogenetik und anderer Gruppierungswissenschaften. *Handbuch der vergleichenden Anatomie der Wirbeltiere* (ed. by Bolk, L., Goppaert, E., Kallius, E. & Lubosch, W.), Band 9, pp. 942–1053. Berlin & Wien.

Zunino, M. (1992). Per rileggere Croizat. *Biogeographia,* 16, 11–23.

2 Common Cause and Historical Biogeography

Lynne R. Parenti

ABSTRACT

The search for a general pattern of area relationships for a group of taxa living in the same region lies at the core of historical or cladistic biogeography. Once a general pattern is discovered, its cause may be inferred. This sequence, first discovering a pattern, then inferring its cause, is the logical foundation of cladistic biogeography. Ignoring this sequence, for example, inferring the causes of a distribution pattern and then using those inferences to reject, or to fail to look for, a pattern, is a rejection of the principles of cladistic biogeography. One reason for ignoring this sequence is misapplication of the principles of phylogenetic systematics in biogeographic analysis. A second reason is inferring *a priori,* using fossil evidence, molecular data, or both, that a taxon is too young or too old to be part of a general pattern and, hence, could not have been affected by the same Earth history events as other taxa. A third reason is assuming that taxa with different dispersal abilities or ecologies are unlikely to be part of the same general pattern, therefore, unlikely to share a distributional history. There are other reasons, many discipline-specific and often correlated with adherence to a center of origin concept. Formation of redundant distribution patterns, with widespread taxa, pruned by extinction, has left some baffling, sometimes seemingly contradictory, distribution patterns worldwide and throughout geologic time. The relationship between Earth history (geology) and phylogeny and distribution patterns (biology) is still under debate. Separating identification of a pattern from inferences about the processes or mechanism that caused it remains a challenge for historical biogeography.

INTRODUCTION

In 1979, the American Museum of Natural History in New York hosted a symposium on *Vicariance Biogeography: A Critique* to "provide a diversified evaluation of the methods and concepts of vicariance biogeography by comparison with other biogeographic concepts such as refuge theory, equilibrium theory, progression rule theory and dispersalism," quoting from the symposium brochure. Gareth Nelson and Donn E. Rosen, then colleagues in the Department of Ichthyology and leaders in the development of vicariance biogeographic theory, methods, and application, were among the symposium organizers and edited the published proceedings

(Nelson and Rosen, 1981). In the Introduction, Rosen, in Nelson and Rosen (1981, p. 5), wrote "It seems doubtful, at least to me, that the notion of congruence between geographical and biological patterns could ever be formulated in a testable way without cladistic information, just as there seems here the promise that the study of cladistic congruence between the earth and its life will be the next 'revolution in the earth sciences' — an integrated natural history of the geological and biological systems." Paleoichthyologist Colin Patterson, of the Natural History Museum, London, a close colleague of both Nelson and Rosen (e.g., Patterson, 1975), also expressed this view (Patterson, 1981).

Donn Rosen died in 1986 at age 57 (see Nelson et al., 1987), just one year before John C. Avise and colleagues (1987) introduced the term "phylogeography" to describe the geographic pattern displayed by intraspecific gene trees — the mitochondrial DNA "bridge" between population genetics and systematics. The subsequent explosion in application of molecular techniques to systematics at the population level and above, as summarized in Avise (2000) and elsewhere, is well known. Application of these techniques to systematics and biogeography has been, on the one hand, invigorating, as it has generated many novel hypotheses of phylogenetic relationships, yet at the same time, detrimental, as it has revived untestable hypotheses of center of origin, recognition of ancestors, and dismissal of the importance of Earth history at all levels, not just plate tectonics, in biogeography (see Ebach, 1999; Humphries, 2004; Nelson, 2004; Parenti & Humphries, 2004, Heads, 2005a; and below).

In the *New York Times* report on the 1979 symposium (Webster, 1979), Niles Eldredge, an AMNH curator of invertebrate paleontology, expressed the view of many in attendance: "... the debate here is whether a certain pattern of species distribution was caused by dispersal or vicariance." The importance of process or mechanism — debating between vicariance and dispersal — was fi ed in biogeographical thought and remains so today for many biogeographers despite the earlier arguments by Platnick & Nelson (1978), more recent statements by Heads (2005a, b), and many others in between, that this diverts attention away from the principal goal of discovering patterns. McLennan and Brooks (2002, p. 1055), for example, note that over the past 30 years, "... the discipline of historical biogeography has diverged into at least two research programmes. These programmes differ fundamentally in their views about the way biological diversity has been produced ...". Cook and Crisp (2005, p. 744) agree: "Biogeographical problems are concerned with process, typically involving discrimination between vicariance and dispersal ...".

Rosen's anticipated "revolution in the earth sciences" — an integrated natural history of the geological and biological systems — has been realized only to a limited degree (e.g., Ebach & Humphries, 2002). It has been stalled not only by some applications of molecular techniques, but also by an emphasis on inferring mechanisms of biological distributions before, or instead of, discovering shared patterns of biological and geological diversity, an emphasis on centers of origins. Further, it has been stalled by adherence to traditional biogeographic regions or realms when global distribution patterns recognized for over 150 years have supported a drastic overhaul or revision of those regions (e.g., Morrone, 2002). In the language of Ebach and Humphries (2002), the revolution in biogeography has been stalled by attempts to generate explanations rather than to discover patterns.

CLADISTIC VS. PHYLOGENETIC BIOGEOGRAPHY

Historical biogeography aims to discover shared patterns of distribution among organisms in a biota, rather than explaining distributions one organism at a time. The search for a general pattern of area relationships as specified by the phylogenetic relationships of taxa living in the same region lies at the core of cladistic biogeography. Once a general pattern is discovered, its cause may be inferred. This sequence, first discovering a pattern, then inferring its cause, is the logical foundation of cladistic biogeography (viz. Platnick & Nelson, 1978; Nelson & Platnick, 1981; Humphries & Parenti, 1986, 1999). Ignoring this sequence, for example, inferring the causes of the distribution of taxa in a biota and then using those inferences to reject, or to fail to look for, a pattern, is a rejection of the principles of cladistic biogeography. Chris Humphries and I (Humphries & Parenti, 1986, 1999) deliberately chose "cladistic biogeography" to identify this research program to emphasize discovery of patterns rather than "vicariance biogeography," which implied an emphasis on identification of Earth history events to explain all distribution patterns, or "phylogenetic biogeography" (see Brundin, 1981; Brooks, 2004), which was tied to center of origin and dispersal concepts, or "panbiogeography" (see Craw et al., 1999; Croizat, 1982), which did not integrate phylogenetic relationships. We did not argue that Earth history had no role in forming the patterns, but that hypothesizing a particular mechanism was secondary to discovering a pattern (see also Ebach, et al., 2003).

Phylogenetic biogeography of Brundin (1966, 1972, 1981) and Hennig (1966) revolved around the notions of a center of origin and dispersal away from such a center by progressively more derived taxa (Figure 2.1a). This method, incorporating the deviation and progression rules (Hennig, 1966), persists (e.g., Bremer, 1992, 1995; Briggs, 1992, 1999; Brooks, 2004, among many others) and dominates phylogeographic studies, as it may be applied to one taxon at a time.

Identification of centers of origin was soundly rejected by Croizat et al. (1974) as untestable; biogeography is a science based on the proposal of hypotheses of distribution that are tested with additional distribution patterns. In his early writings on historical biogeography, however, Nelson (1969, p. 687) advocated an explicit, untested notion of dispersal: "assuming the relationships [of Figure 2.1b] who could doubt, e.g., that the occurrence of this lineage in area x is a secondary and a relatively late one?" The conclusion of dispersal was made as an observation, not a hypothesis open to test. Five years later, Nelson (1974, Figure 1) recanted in his alternative formalization of historical biogeography, arguing that there was no such evidence for dispersal and that the ancestral area could be estimated by adding together the areas of the descendant taxa. The method described briefly by Nelson (1969) persists, however, and has even been named, cladistic subordinateness (Enghoff, 1993), to denote presumed evidence of secondary dispersal by a taxon with a restricted distribution nested among more broadly distributed derived taxa. Nelson and Ladiges (2001, Figure 1c) reworked Nelson's (1974, Figure 1) example, to demonstrate that

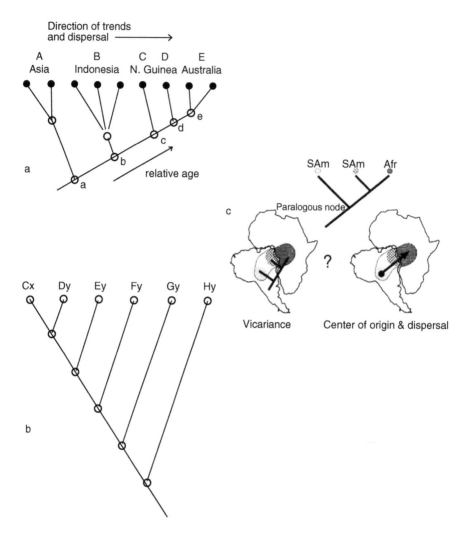

FIGURE 2.1 (a) Phylogenetic biogeographic hypothesis specifying sequential dispersal of increasingly derived taxa away from a center of origin, from Brundin (1972, Figure 3); (b) historical biogeographic hypothesis specifying the dispersal of taxon C from area y to area x, from Nelson (1969, fig, 3); (c) alternative interpretations of a distribution pattern from Nelson and Ladiges (2001, Figure 10), following Nelson (1974).

either a vicariant or dispersal hypothesis may be used to interpret the distribution pattern — in other words, that the paralogous node in Figure 2.1c, repetition of South America, adds no information to area relationships and does not specify a center of origin in South America.

Another kind of center of origin interpretation of single cladograms — optimization of habitat — was made by Rosen (1974) in his study of salmoniform phylogeny and biogeography. The phylogenetic hypothesis of salmoniform fishe (Figure 2.2a) was interpreted as follows (Rosen, 1974, p. 313), "... salmoniforms

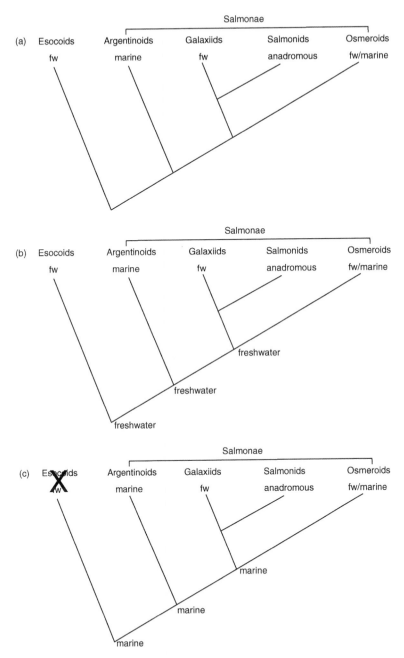

FIGURE 2.2 (a) Cladogram of relationships of the subgroups of the order Salmoniformes (sensu Rosen, 1974, Figure 39), including basic habitat; fw = freshwater; galaxiids are diadromous, but were treated as freshwater by Rosen (1974) because they breed in freshwater. (b) Habitats optimized according to Rosen's (1974) criteria given a freshwater sister group of the Salmonae. (c) Alternate optimization of habitats following Rosen's (1974) hypothetical proposal of a marine sister group of the Salmonae. (d) Reconstruction of ancestral habitat of marine and freshwater for Salmonae.

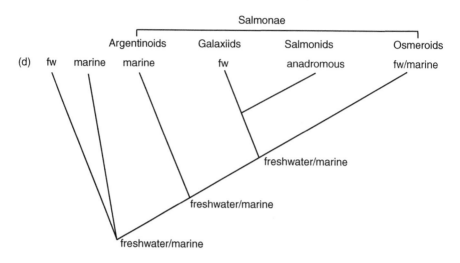

FIGURE 2.2 (Continued)

are therefore primitively freshwater fishes that have twice become secondarily marine … . It should be noted, however, that if esocoids are incorrectly associated in this order and the sister group of Salmonae should prove to be marine instead of freshwater, then the Salmonoidei may be secondarily anadromous [migrating between marine and freshwaters] … ." Here, Rosen (1974, Figure 2.2b and 2.2c) treated a cladogram as a phylogenetic tree and habitat (marine vs. freshwater) as a character to be optimized. An alternate interpretation (Figure 2.2d) would entail reconstructing a widespread ancestral habitat for the Salmonae — both freshwater and marine — rather than relying on the present-day habitat of the hypothesized living sister-group to interpret historical habitat distribution (see also Parenti, submitted). Coincidentally, McDowall (1997) also concluded that diadromy is an ancestral condition for the Salmonidae, Osmeridae, and Galaxiidae (Parenti, in press).

Erasing the notion of centers of origin from biogeography became a goal of Nelson and Rosen once teamed with Croizat (e.g., Croizat et al., 1974). Despite explicit, repeated criticism that optimization of nodes in biogeography implies a center of origin hypothesis (e.g., Nelson, 1974, 2004; Platnick & Nelson, 1978; Parenti, 1981, 1991; Parenti & Humphries, 2004), it persists as a method among both neontologists and paleontologists, especially to infer an ancestral habitat (e.g., Brooks & McLennan, 1991; Stearley, 1992; Winterbottom & McLennan, 1993; McLennan, 1994; Lovejoy & Collette, 2001; Thacker & Hardman, 2005, among many others).

METHODS OF HISTORICAL BIOGEOGRAPHY

Methods of historical or cladistic biogeography were developed to interpret the area relationships of three or more taxa in a biota, using the area cladogram, or areagram, in which the names of taxa were replaced by the name of the areas in which they live (e.g., Brundin, 1972; Rosen, 1978). Component analysis was developed by

Nelson and Platnick (1981; Platnick & Nelson, 1978) and programmed by Page (1989, 1993; see also Page, 1990, Morrone & Carpenter, 1994; Crisci et al., 2003). Brooks parsimony analysis (BPA) was developed by Brooks (1981) for host–parasite relationships, following Hennig's (1966) parasitological method, and expanded to encompass a range of historical biogeographic questions by Wiley (1987, 1988), Brooks & McLennan (1991, 2002), Lieberman (2000, 2004), Van Veller et al., (2000) and Van Veller & Brooks (2001), among others. Both methods adopt the simple premise that organism relationships reflect their distributional history, complex in practice because organisms and areas do not act as one, especially through geological time (see discussions in Grande, 1985; Cracraft, 1988; Hovenkamp, 1997). There are other methods of historical biogeographic analysis (see Crisci et al., 2003), but it is not my purpose to review all of them here. Widespread taxa (taxa living in more than one area), extinction (taxa missing from areas) and redundant distributions wreak havoc on patterns of areagrams (see Platnick & Nelson, 1978; Nelson & Platnick, 1981; Nelson & Ladiges, 1996). Widespread taxa, missing areas and redundant distributions are analyzed and interpreted differently by component analysis and BPA, and therein lies the basis of a quarter century of debate (compare Ebach et al., 2003 with McLennan & Brooks, 2002, for example). A taxon that lives in two areas, A and B, is interpreted by BPA as indicating that areas A and B share a closer history than either does to another area, say area C. Under component analysis, presence of a taxon in areas A and B does not preclude that either one of those areas shares a closer history with area C. A widespread taxon could have failed to differentiate as the areas it occupied (A and B) diverged (Nelson & Platnick, 1981).

Donn Rosen would have had methods develop in another way, with biologists collaborating with geologists, as he did in formulating his hypotheses of Caribbean historical biogeography (Rosen, 1985, see below). I was privileged to be one of Rosen's graduate students at the American Museum of Natural History from 1975 to 1980. Naturally, I sent him a draft of the proposal for the first edition of *Cladistic Biogeography* (Humphries & Parenti, 1986) for comment. He replied:

I think your and Chris' book outline is fine so far as it goes, but that it lacks a major ingredient — perhaps the most important one if we're concerned with the future development of biogeography. This ingredient is a discussion of cladistic historical geology, without which 'the evolution of the earth and its life' is a pipe dream. Historical geology *is* analyzable cladistically, and both theoretical *and real* examples must be presented to explain to geologists how to go about providing the independent data base necessary to demonstrate consilience. I've been consulting recently with knowledgeable geologists and will try to put together a method for dealing with the geology. It's more complicated than I thought it would be, but still workable. My suggestion, therefore, is that you and Chris gather up some of the stray geologists in Britain who can give you some well-documented histories and convert these into cladistic patterns that might be compared with some good biological ones. This is the way to convince geologists that they have much to learn from biology and that they had best set to work producing supportable cladograms. The other important feature of such a chapter is that it would describe a common language for biologists and geologists that would enhance a needed exchange of ideas now sorely lacking. It might even provoke a methodological revolution in historical geology (Donn E. Rosen, *in litt.*, 3 March 1982; italics in original).

Geological cladograms have not been pursued with the same interest and enthusiasm as areagrams derived from taxon cladograms, for better or worse. Better because we, as biologists, are not tied to consensus interpretations of geological hierarchies (see McCarthy, 2003, 2005, for example). Worse because without independent geological cladograms, biologists rely on prevailing geological hierarchies as the best interpretations of Earth history, as in event-based methods (e.g., Sanmartín & Ronquist, 2004), which therefore cannot discover patterns that conflict with current geology (see Heads, 2005a, b). Geologists have provided detailed summaries of the geological histories of complex regions, but not explicit geological cladograms. The detailed reconstructions of Southeast Asian geology over the past 55 million years by Robert Hall, Department of Geology, University of London, and colleagues are notable in this regard (e.g., Hall, 1996, 2002; http://www.gl.rhul.ac.uk/ seasia/).

In developing methods of historical biogeography using areagrams, the principle of parsimony has been applied to choose among competing hypotheses. The analogy between taxa and characters in phylogenetics and areas and taxa in biogeography is not precise, however, and efforts to make it so have not addressed adequately the above-mentioned differences between the goals of component analysis and BPA with respect to widespread taxa (viz. Kluge, 1988; Wiley, 1988; Brooks & McLennan, 1991; Lieberman & Eldredge, 1996). Further, Nelson & Ladiges (1991, 1993) have developed an areagram-based method termed three-area statements analysis to extract informative statements about area relationships from complex areagrams that are characterized by duplicated or paralogous nodes. Most parsimonious interpretations of area/taxon matrices, as required by BPA, do not necessarily account for geographic paralogy, area hierarchies that are interpreted as contrary to a general pattern because of duplication or overlap in geographic distribution of taxa (Nelson & Ladiges, 1996). The differences in these general classes of methods have been characterized by McLennan & Brooks (2002, pp. 1056–1057) and Brooks (2004) as an emphasis on area relationships (component analysis) vs. speciation processes (BPA). The so-called non-vicariant (i.e., non-Earth history) elements of diversity listed by McLennan & Brooks (2002, p. 1063) as explanations for a pattern such as non-response to a vicariant event (widespread taxa), extinction and lineage duplication were all recognized by Nelson & Platnick (1981), Nelson & Ladiges (1996) and others as common phenomena that all biogeographic methods should address. The two other non-vicariant elements of diversity listed by McLennan & Brooks (2002), peripheral isolates speciation and post-speciation dispersal, are hypotheses that, like all hypotheses, are open to test. They are interpretations, not observations. Proposing them does not necessarily refute allopatric speciation or vicariance (see also Ebach et al., 2003), although it does endorse the philosophy that one aim of historical biogeography is to generate an explanation as well as to discover a pattern.

GEOLOGY AND DISPERSAL

Cladistic biogeographers have invoked Earth history explanations for biotic distribution because of the overwhelming congruence between the two, on both local and global scales (e.g., Rosen, 1974, 1976, 1978, 1979, 1985), for both freshwater (e.g., Sparks & Smith, 2005) and marine taxa (e.g., Springer, 1982) and over geological time periods (e.g., Upchurch, et al., 2002). Such interpretations have been viewed repeatedly as a neglect of, or disdain for, the phenomenon of dispersal (recently, De Queiroz, 2005). By 1978, Rosen (1978, p. 750) had dispensed with this argument in a footnote. Nonetheless, it persists, for example, in BPA coding of matrices for the analysis of "geodispersal".

"Geodispersal" is a term, coined by Lieberman & Eldredge (1996), that refers to (Lieberman, 2000, p. 22) "… tectonically mediated events such as continental collision [that] can … bring formerly separated faunas into contact with one another producing a pattern of congruent range expansion in several groups." As a concept, it has been around at least since the early writings of Alfred Russel Wallace. In 1863, Wallace read a paper at a meeting of the Royal Geographical Society of London, *On the Physical Geography of the Malay Archipelago* (Wallace, 1863). Although Wallace's understanding of Indonesian geology was rudimentary by today's standards, and he may not have conceived of areagrams, he understood the potential for mixing of biotas on either side of what was to become known as Wallace's Line following geological rearrangement and proposed it as a mechanism to produce composite biotas in the Philippines and Sulawesi (Wallace, 1863, pp. 232–233): "The nature of the contrast between these two great divisions of the Malay Archipelago will be best understood by considering what would take place if any two of the primary divisions of the earth were brought into equally close contact … . Some portion of the upraised land [between two continents] might at different times have had a temporary connection with both continents and would then contain a certain amount of mixture in its living inhabitants."

Nonetheless, the idea of mixing biotas following geological rearrangement was considered novel by Lieberman (2000) because he interpreted it as having been rejected by cladistic or vicariance biogeographers: "Originally, it was believed that biogeographic patterns in different groups of organisms could only result from vicariance. However … in addition to vicariance, there is geodispersal. In vicariance, congruence results from the formation of geographic barriers, while in geodispersal it is produced when geographic barriers fall" Lieberman (2000, p. 186).

"Originally" I interpret to mean in the early writings of Nelson, Platnick, and Rosen, rather than during the time of Wallace (viz. Humphries, 2004). But, these three "original" vicariance biogeographers had an explicit concept of mixing of biotas following geologic rearrangement. In what could be interpreted as an effort to implement the "… next 'revolution in the earth sciences' — an integrated natural history of the geological and biological systems", Rosen (1985, pp. 652–653) discussed the history of the Caribbean biota with regard to the potential complexity of biogeographic patterns that would result from accretion and fragmentation. "Biotic mixing" here is dispersal of members of one biota into the range of another following suturing of the respective land masses, or what would later be called "geodispersal":

> … in a general cladogram of areas based on both fragmentation and accretion, land hybridizations would be represented either as reticulations … or, as in the case of biological hybridizations, as unresolved branches representing the parent taxa and the hybrid … One might suppose that land hybridization … would lead to a sharing of biotas by the joined fragments, but that supposition requires the subsidiary idea, probably correct in some instances but not others, that the suture zone represents no obstacle to biotic mixing."

Lieberman (2000, p. 83) claimed that Platnick and Nelson were opposed to dispersal in the form of range expansion as an explanation for a biogeographic pattern: "Platnick and Nelson (1978, p. 7) also invoked a geological example that produced what I have defined as geodispersal — the collision between India and Australia during the Cenozoic era — and they termed it 'biotic-dispersal.' I consider this term inaccurate and potentially confusing because the range expansion is not caused by biological factors at all ... rather it is related to geological change. Therefore, I prefer the term geodispersal."

He continued (Lieberman, 2000, p. 83): "... what is important is that the existence of the *same phenomenon* [my italics], congruent episodes of range expansion was recognized by Platnick and Nelson (1978), two of the primary architects of the vicariance biogeography approach, although they were generally virulently opposed to invoking traditional dispersal in any explanation whatsoever ... ." Here, Platnick and Nelson's understanding of biotic-dispersal, a senior synonym of geodispersal in taxonomic parlance, is dismissed by the claim that they opposed dispersal in any form. Not so, as Platnick & Nelson (1978, p. 8) were not opposed to a notion of dispersal, but argued that it is the discovery of a pattern, shared area relationships among taxa, that is primary in historical biogeography and an explanation for a pattern, breaking apart of land masses, extinction, dispersal, and so on, that is secondary to that discovery of pattern (see Crother, 2002). Geological rearrangement resulting in biotic mixing has been proposed as an explanation for complex, composite biotas at least since Wallace (1863; see e.g., Carpenter, 1998; Carpenter & Springer, 2005). Integration of non-Hawaiian islands into the Hawaiian Islands was proposed as a mechanism for the mixing of two different biotas and as an explanation for endemism in the Hawaiian Islands (Rotondo, et al., 1981).

Interpreting evolution of the Caribbean biota required complex, reticulate areagrams, according to Rosen (1985), to incorporate adequately the complex patterns over geologic periods. Concurrently, a time control refinement for cladistic biogeography was proposed by Grande (1985) as a method to identify historical biogeographic patterns for fossil biotas from specified geologic periods. This was in response to his conclusion that summarizing complex patterns of vicariance, accretion, and dispersal for a biota in a single areagram would require the cladogram to be either reticulate or unresolved (Grande, 1985, p. 238): "The *Recent* [italics in original] biota of certain geographic areas could be the biota least likely to show a fully resolved biogeographic pattern for a given area because it may contain descendants of older endemic components ... *plus* species ... which later dispersed into the area or were introduced ... *minus* all those species which have become extinct." For the North American fish biota, Grande (1985) concluded that fi e Green River Eocene taxa have trans-Pacific relationships, whereas this affinity is masked among Recent taxa because of the presence of additional taxa with trans-Atlantic affinitie and because of extinction of Australasian taxa. He argued that a series of areagrams, each representing a different time period, would eliminate the conflict that characterizes a single, summary cladogram. Older and younger components to general areagrams have also been dealt with by recognizing that composite areas, for example, South America, could be part of two patterns of possibly different ages (see Parenti, 1981; Humphries & Parenti, 1986, 1999; Wanntorp and Wanntorp, 2003) despite statements to the contrary (e.g., Donoghue & Moore, 2003).

Only BPA can reconcile the differences among areagrams over geologic time periods, however, according to Lieberman (2000, 2004). In implementations of the method, geodispersal is coded as an ordered, multistate character. Range expansion of a taxon from areas A, B, and C to areas A, B, C, and D, for example, is coded as state 1 in areas A through C and state 2 in area D. The difficulty with such *a priori* coding to implement geodispersal is that the distribution of a single taxon is used to infer the mechanism or process by which it became distributed. That is, a taxon found in areas A through C at one time period that is found in areas A through D at a later time period is interpreted as having expanded its range. This might be true, but an alternate interpretation, for example, that the taxon lived in area D during the earlier time period but has not been found in that epoch, cannot be rejected. Lieberman (2000, p. 148) acknowledges this problem and appeals to completeness of the fossil record. Nonetheless, coding the state as ordered specifies a direction or episode of dispersal — an extension of Hennig's (1966) and Brundin's (1966) progression rule — rather than asks if that inferred dispersal is a hypothesis supported by the distributions of the taxa in the biota.

Coding for geodispersal also does not take into account that some nodes may be paralogous. Paralogy or redundancy in an areagram may be caused by (Nelson & Ladiges, 1996, pp. 11–12): "... tectonics, dispersal, sympatric speciation, mistaken relationships among organisms, imprecise characterizations of geographic areas, and so on." The relationship of a paralogous area conflicts with itself because of overlap or redundancy; sympatric areas are paralogous. For example, an area A may in one portion of an areagram be related to area B, and in another portion of the cladogram be related to area C because of redundancy and extinction. Neither set of area relationships, (A,B) or (A,C), conflicts with a general pattern of (((A,B),C),D), as (A,C) may result from redundancy and extinction in one portion of the biota.

Eliminating paralogous nodes identifies informative statements of area relationship. Eliminating uninformative data has been interpreted as violating the principle of parsimony and therefore as scientifically invalid (see Van Veller & Brooks, 2001). Yet, if a widespread taxon in areas A and C, above, does not mean necessarily that areas A and C share a close history, then it matters little whether there are 1, 10, or 1000 such distributions; each is equally uninformative. Interpretation of informative data, not misapplication of parsimony, is the issue. I offer another analogy. Using the number of species in a monophyletic lineage to support a biogeographic hypothesis about that lineage is analogous to using the number of specimens examined to support a phylogenetic hypothesis of that lineage. That is, coding a morphological character from one specimen or one hundred does not affect the coding of the character in a phylogenetic analysis; it may give one more confidence in the results, but no more explanatory power.

MOLECULES AND TIME

Common causes of congruent distribution patterns have been rejected outright because of the assumption that one or another taxon is too young to have been affected by Earth history events proposed to have affected another taxon (e.g., Lundberg, 1993, for fishes, Voelker, 1999, for birds; De Queiroz, 2005, for oceanic organisms; see Nelson & Ladiges, 2001 for a historical review). Many of the reasons why it would be folly to ignore a wealth of biogeographic congruence in favor of dispersal and assume molecular data suggest the relative youth of many taxa, as advocated by De Queiroz (2005), are detailed by Heads (2005a) in a cogent review. I comment here on two applications of molecular data to historical biogeography, both from the marine realm, to warn against a rejection of biogeographic patterns using estimates of divergence times based on molecular data.

For some 60 years, coelacanths, basal sarcopterygian fishes, were believed to be represented by just one living species, the western Indian Ocean *Latimeria chalumnae* Smith, 1939. In 1997, a second population was discovered off the coast of Sulawesi, Indonesia, in the western Pacific, and described as a new species, *Latimeria menadoensis* Pouyaud *et al.* (1999). Systematic ichthyologists, a paleontologist (Forey, 1998) and a neontologist (Springer, 1999), predicted that the two populations would be found to be distinct based on several factors, including the widespread distribution of the populations, present-day ocean current patterns, the narrow habitat of the deep-sea, lava-tube dwelling coelacanth and the prevailing understanding of the geological history of the Indo-Pacifi region. At first, independent studies using partial mitochondrial gene sequences estimated that the two species diverged from each other between 6.3 and 4.7 Mya (Holder et al., 1999) or as recently as 1.3 mya (Pouyaud et al., 1999). The latter authors equivocated about the type and amount of molecular differentiation between the two populations, but nonetheless described the Sulawesi population as a new species, effectively proposing a hypothesis that the differences between the two populations would be corroborated. A third estimate of divergence time of 40 to 30 mya, based on whole mitochondrial genomes (Inoue et al., 2005) is more in line with Springer's (1999) hypothesis that the collision of India with Eurasia 50 mya bisected the widespread, ancestral range of the coelacanth and, therefore, with the hypothesis that there are two living species. Coelacanth species may be younger or older than the inferred collision of India with Eurasia, or that event may not have caused separation of the ancestral population (viz, Heads, 2005a, b). Estimate of the species' divergence times compared with an estimate of the timing of a major geological event, however, may be used to formulate a hypothesis of the evolution of living coelacanths that incorporates our best, current understanding of speciation and of geology. That there are three different estimates of divergence times based on the application of molecular sequence data is but one indication of how questionable still are these methods, in large part because of their reliance on fossils for calibration (see also Graur & Martin, 2004).

Astralium rhodostomum (Lamarck) is a wide-ranging, reef-dwelling, Indo-west Pacific turbinid gastropod with a short planktonic larval stage (Meyer et al., 2005). Mitochondrial DNA evidence supports an interpretation that the species complex comprises two widespread, overlapping, deeply divergent clades with a high degree of endemism. There are hypothesized to be at least 30 geographically isolated clades throughout islands and island groups ranging from Phuket and Cocos Keeling in the Indian Ocean to Rapa in the western Pacific (Meyer et al., 2005, Figure 1). This high level of diversity and endemism has been viewed as expected for terrestrial species throughout the region, but not for marine taxa which are generally interpreted as having wide-ranging species throughout the Indo-west Pacific that have been distributed by long-range dispersal (Meyer et al., 2005). Notable exceptions to this interpretation come from Springer

(1982), who documented the coincidence of species limits with tectonic plate boundaries for fishes and select other marine taxa, and Gill (1999) and Gill and Kemp (2002), who hypothesized that many currently recognized widespread marine fish species comprise two or more allopatric, cryptic species with endemic distributions.

The two clades of *Astralium rhodostomum,* Clade A and Clade B (Meyer et al., 2005, Figure 3), each repeat in four areas: Ryukyu, New Caledonia, Marianas, and Tonga, Samoa, Niue, and Fiji. The last group, Tonga, Samoa, Niue and Fiji are sister areas in both clades and therefore treated as one area here. The areagram for Clade A is reproduced, along with that portion of the areagram for Clade B that includes the repeated areas (Figure 2.3a and 2.3b).

Each areagram is completely resolved. The shared information in each can be extracted easily either through reduced area-cladograms or three-area statements analysis (Nelson & Ladiges, 1996). There are four three-area statements specifie by the areagram for Clade A, and three three-area statements for Clade B (Figure 2.3c and 2.3d). There is one three-area statement common to both (Figure 2.3d), which specifies a set of area relationships that the two clades share. Therefore, not only is there a high degree of endemism in the overlapping clades, but a repeated, inferred shared, history. Molecular data have the potential to unravel additional fine scale patterns of relationship among endemic areas, especially throughout the marine realm where they have not been expected.

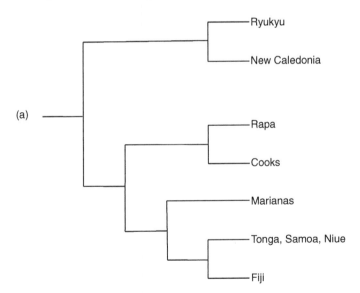

FIGURE 2.3 (a) *Astralium rhodostomum,* Clade A areagram (following Meyer, Geller, and Paulay, 2005, Figure 3). (b) Clade B, groups B1 and B2 areagram. "Philippines" comprises Palawan and Central Visayas; "Eastern I-A Archipelago" comprises: Papua New Guinea/Queensland, Sulawesi, Manus, and New Britain. (c) The four three-area statements specified by the areagram for Clade A. (d) The three three-area statements specified by the areagram for Clade B, groups B1 and B2. There is one shared three-area statement, starred and shaded in Figure 2.3c and 2.3d.

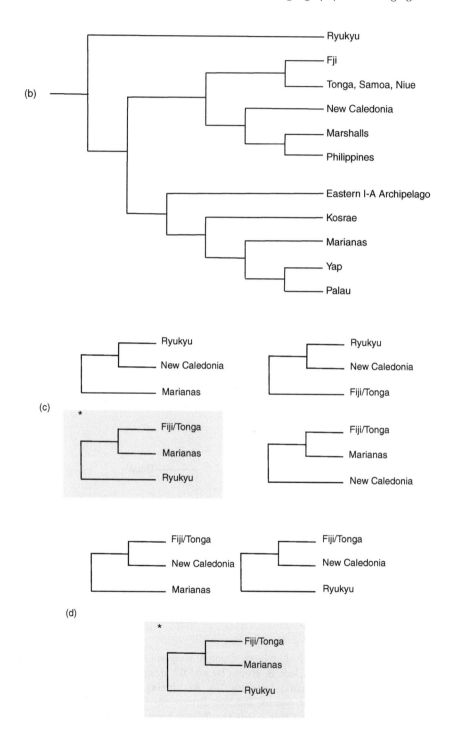

FIGURE 2.3 (Continued)

GLOBAL BIOGEOGRAPHIC PATTERNS VS.
BIOGEOGRAPHIC REALMS OR REGIONS

Nearly 150 years ago, Sclater (1858) divided the world into six terrestrial biogeographic regions to describe the distribution of birds: Nearctic, Palaearctic, Neotropical, Ethiopian, Oriental, and Australian (see Humphries & Parenti, 1999, Figure 1.9). These were considered useful by Wallace (1876) to describe the world's fauna and have been so used ever since, despite the fact that it has been known for as long that they omit some of the most distinctive and intriguing patterns of plant and animal distributions: pantropical and antitropical distributions. In 1866, Murray published a lavishly illustrated monograph on the distribution of the world's mammal fauna (Murray, 1866). Murray's map 53 colorfully outlines the antitropical distribution of whale bone whales, map 51 the pantropical distribution of Sirenia. South and Central America has been considered nearly synonymous with the Neotropical region despite its composite biota, part tropical and part austral, part Ostariophyan and part austral Salmoniformes, according to Rosen (1974, Figure 45). Repeatedly, historical biogeographic analyses recognize the "two South Americas" (Parenti, 1981; Humphries & Parenti, 1986, 1999; Carpenter, 1993; Arratia, 1997; Vari & Weitzman, 1990), and Nelson & Ladiges (2001) imply "three South Americas": trans-Pacific, trans-Atlantic, and austral. Further, the unjustified separation of marine and freshwater taxa in biogeography has lead to description of at least eight marine regions in addition to Sclater's six classic regions (see Mooi & Gill, 2002). A reclassification of global biogeographic regions is long overdue.

Panbiogeographers (Croizat, 1958, 1964; Craw et al., 1999) have argued for a classification of the world's biotic regions centered on ocean basins, not continents, to reflect the composite nature of continental biotas. Other biogeographers have agreed (e.g., Parenti, 1991; Morrone, 2002; Humphries, 2004). The relationships among these areas has been interpreted by Humphries and Parenti (1986, 1999) and Nelson and Ladiges (2001, Figure 8) who proposed a cladistic summary of Croizat's (1958, Figure 4a) and Craw et al.'s (1999) global distribution patterns. These patterns endlessly repeat, and extinction has pruned them (Figure 2.4b). Nonetheless, if they are to be useful, patterns must have explanatory and predictive power. This may be demonstrated for the atherinomorph Scomberesocidae, a family of epipelagic marine fishes distributed broadly in temperate and tropical oceans (Hubbs & Wisner, 1980). Scomberesocids have been classified in four genera, *Scomberesox* and its miniature sister genus, *Nanichthys,* and *Cololabis* and its miniature sister genus, *Elassichthys.* Their relationships and distribution pattern (Figure 2.5) reflect the global pattern (Figure 2.4a), with some pruning. *Nanichthys*, in the tropical Atlantic and Indian oceans is sister to *Scomberesox* in the North Atlantic and in austral seas. *Cololabis* is boreal in the northern Pacific; *Elassichthys* is tropical in the eastern central Pacific Such patterns can be used to make predictions about where closely related taxa may be found: 'is there an austral *Cololabis?*', for example.

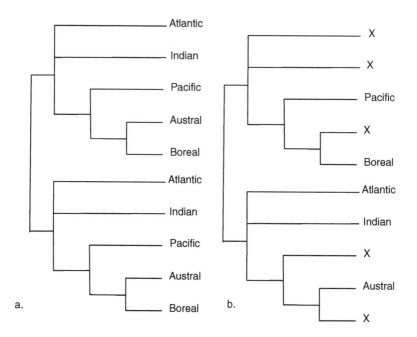

FIGURE 2.4 (a) Cladistic summary of Croizat's (1958) and Craw, Grehan, and Heads's (1999) global distribution patterns, from Nelson and Ladiges (2001, Figure 8), drawn to demonstrate repetition. (b) The pattern of 2.4b pruned by hypothesized extinction events marked by an X.

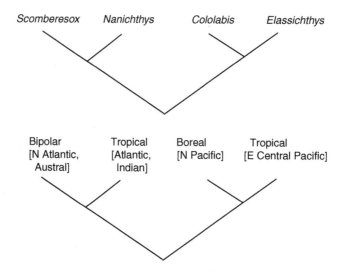

FIGURE 2.5 Classification of the genera of the epipelagic marine atherinomorph fish family Scomberesocidae, following Hubbs and Wisner (1980), expressed in a taxon cladogram, above, and areagram, below.

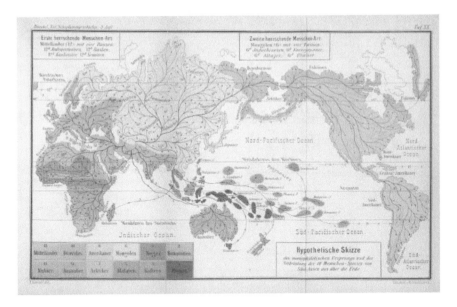

COLOR FIGURE 1.8 Reproduction of *Hypotheische Skizze des monophyletischen Urs-prungs und der Verbreitung der 12 Menschen-Species von Lemurien aus üe r die Erde* from the 8th edition of *Natü liche Schpf ungsgeschichte* (Haeckel, 1889, taf. 20).

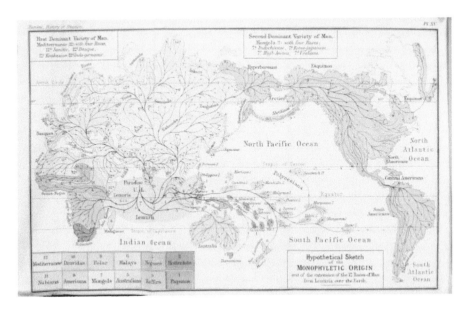

COLOR FIGURE 1.9 Reproduction of *Hypothetical Sketch of the Monophyletic Origin and the Extension of the 12 Races of Man from Lemuria over the Earth* (Haeckel, 1876a, taf. XV).

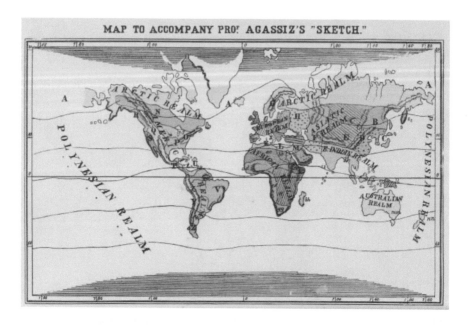

COLOR FIGURE 1.10 Agassiz's map of realms harbouring collections of organisms unique to each region (Agassiz, 1854, facing p. lxxviii).

CONCLUSIONS

Incorporating center of origin explanations into the methods of historical biogeography has resulted in the generation of untestable hypotheses rather than the discovery of testable patterns. The principle of a center of origin was part of the early formalizations and implementations of historical biogeography (e.g., Nelson, 1969), but subsequently rejected (Croizat et al., 1974; Nelson, 1974). Discovery of patterns is primary in biogeography — identifying the mechanism by which the pattern was produced is secondary. This was stated clearly in early formalizations of vicariance biogeography (e.g., Platnick & Nelson, 1978), but it has been misinterpreted or ignored. These and many of the other papers cited here are reproduced in the immensely valuable volume by Lomolino *et al.* (2004), which makes them more accessible and available for study and interpretation of methodological developments.

Many intriguing biogeographic patterns remain to be discovered based on analysis of phylogenetic relationships using morphology and molecules, on specimens collected through field exploration, or on revision of museum and herbarium collections. Molecular estimates of divergence times are as yet unable to reject the ancient differentiation of lineages as predicted from biogeographic patterns. Recognition of a high degree of endemism among marine populations is one example of how molecular data can be used to demonstrate diversity over broad geographic scales.

Global biogeographic realms/regions that have been used uncritically for over 150 years need to be replaced to recognize realistic, definable, repeatable areas. They should incorporate both terrestrial and marine components, hence, ocean basins (viz. Croizat, 1958). They need to incorporate major biogeographic patterns such as pantropical and antitropical distributions and acknowledge that continental biotas are composite. This will be the start of the "revolution" in biogeography that is long overdue.

ACKNOWLEDGEMENTS

I am grateful to Malte Ebach and Ray Tangney for inviting me to participate in the Fifth Biennial Conference of the Systematics Association where a version of this paper was presented. Chris Humphries and Dave Williams kindly read and commented on a previous manuscript that incorporated some of the ideas expressed here. Gary Nelson provided references.

REFERENCES

Arratia, G. (1997). Brazilian and Austral freshwater fish faunas of South America. A contrast, in tropical biodiversity and systematics. *Proceedings of the International Symposium on Biodiversity and Systematics in Tropical Ecosystems,* Bonn, 1994 (ed. by Ulrich, H.), pp. 179–187. Zoologisches Forschunginstitut und Museum Alexander Koenig, Bonn.

Avise, J.C. (2000). *Phylogeography: The History and Formation of Species.* Harvard University Press, Cambridge, MA.

Avise, J.C., Arnold, J., Ball, R.M., Jr., Bermingham, E., Lamb, T., Neigel, J.E., Reeb, C.A. & Saunders, N.C. (1987). Intraspecific phylogeography: the mitochondrial DNA bridge between population genetics and systematics. *Annual Review of Ecology and Systematics,* 18, 489–522.

Bremer, K. (1992). Ancestral areas: a cladistic reinterpretation of the center of origin concept. *Systematic Biology*, 41, 436–445.

Bremer, K. (1995). Ancestral areas: optimization and probability. *Systematic Biology*, 44, 255–259.

Briggs, J.C. (1992). The marine East Indies: centre of origin? *Global Ecology and Biogeography Letters*, 2, 149–156.

Briggs, J.C. (1999). Extinction and replacement in the Indo-West Pacific Ocean, *Journal of Biogeography*, 26, 777–783.

Brooks, D.R. (1981). Hennigs parasitological method: a proposed solution. *Systematic Zoology*, 30, 229–249.

Brooks, D.R. (2004). Reticulations in historical biogeography: the triumph of time over space in evolution. In *Frontiers of Biogeography. New Directions in the Geography of Nature* (ed. by Lomolino, M.V. & Heaney, L.R.), pp. 125–144. Sinauer Associates, Inc., Sunderland, MA.

Brooks, D.R. & McLennan, D.A. (1991). *Phylogeny, Ecology and Behavior.* University of Chicago Press, Chicago.

Brooks, D.R. & McLennan, D.A. (2002). *The Nature of Diversity.* University of Chicago Press, Chicago.

Brundin, L. (1966). Transantarctic relationships and their significance as evidenced by midges. *Kungliga Svenska Vetenskapsakademiens Handlingar*, 4, 1–472.

Brundin, L. (1972). Phylogenetics and biogeography. *Systematic Zoology*, 21, 69–79.

Brundin, L. (1981). Croizat's panbiogeography versus phylogenetic biogeography. In *Vicariance Biogeography: A Critique* (ed. by Nelson, G. & Rosen, D.E.), pp. 94–158. Columbia University Press, New York.

Carpenter, J.M. (1993). Biogeographic patterns in the Vespidae (Hymenoptera): two views of Africa and South America. In *Biological Relationships Between Africa and South America* (ed. by Goldblatt, P.), pp. 139–155. Yale University Press, New Haven, CT.

Carpenter, K.E. (1998). An introduction to the oceanography, geology, biogeography, and fisheries of the tropical and subtropical western and central Pacific. In *FAO Species Identification Guide for Fishery Purposes. The Living Marine Resources of the Western Central Pacific* (ed. by Carpenter, K.E. & Niem, V.H.), 1–17. FAO, Rome.

Carpenter, K.E. & Springer, V.G. (2005). The center of the center of marine shore fis biodiversity: the Philippine Islands. *Environmental Biology of Fishes*, 72, 467–480.

Cook, L.G. & Crisp, M.D. (2005). Directional asymmetry of long-distance dispersal and colonization could mislead reconstructions of biogeography. *Journal of Biogeography*, 32, 741–754.

Cracraft, J. (1988). Deep-history biogeography: retrieving the historical pattern of evolving continental biotas. *Systematic Zoology*, 37, 221–236.

Craw, R.C., Grehan, J.R. & Heads, M.J. (1999). *Panbiogeography: Tracking the History of Life.* Oxford University Press, Oxford.

Crisci, J.V., Katinas, L. & Posadas, P. (2003). *Historical Biogeography, an Introduction.* Harvard University Press, Cambridge.

Croizat, L. (1958). *Panbiogeography.* Published by the author, Caracas.

Croizat, L. (1964). *Space, Time, Form: the Biological Synthesis.* Published by the author, Caracas.

Croizat, L. (1982). Vicariance, vicariism, panbiogeography, 'vicariance biogeography', etc. a clarification. *Systematic Zoology*, 31, 291–304.

Croizat, L., Nelson, G. & Rosen, D.E. (1974). Centers of origin and related concepts. *Systematic Zoology*, 23, 265–287.

Crother, B.L. (2002). O Patterns! Wherefore art thou patterns? (with apologies to Shakespeare). *Journal of Biogeography*, 29, 1263–1265.

De Queiroz, A. (2005). The resurrection of oceanic dispersal in historical biogeography. *Trends in Ecology and Evolution*, 20, 68–73.

Donoghue, M.J. & Moore, B.R. (2003). Toward an integrative historical biogeography. *Integrative and Comparative Biology*, 43, 261–270.

Ebach, M.C. (1999). Paralogy and the centre of origin concept. *Cladistics*, 15, 387–391.

Ebach, M.C. & Humphries, C.J. (2002). Cladistic biogeography and the art of discovery. *Journal of Biogeography*, 29, 427–444.

Ebach, M.C., Humphries, C.J. & Williams, D.M. (2003). Phylogenetic biogeography deconstructed. *Journal of Biogeography*, 30, 1285–1296.

Enghoff, H. (1993). Phylogenetic biogeography of a Holarctic group; the julidan millipedes. Cladistic subordinateness as an indicator of dispersal. *Journal of Biogeography*, 20, 525–536.

Forey, P. (1998). A home from home for coelacanths. *Nature*, 395, 319–320.

Gill, A.C. (1999). Subspecies, geographic form and wide-spread Indo-Pacific coral-reef fis species: a call for change in taxonomic practice. *Proceedings of the Fifth Indo-Pacific Fish Conference, Nouméa, 1997* (ed. by Seret, B. & Sire, J.-Y.), pp. 79–87. Societé Française d'Ichtyologie, Paris.

Gill, A.C. & Kemp, J.M. (2002). Widespread Indo-Pacific shore-fish species: a challenge for taxonomists, biogeographers, ecologists and fishery and conservation managers. *Environmental Biology of Fishes*, 65,165–174.

Grande, L. (1985). The use of paleontology in systematics and biogeography, and a time control refinement for historical biogeography. *Paleobiology*, 11, 234–243.

Graur, D. & Martin, W. (2004). Reading the entrails of chickens: molecular timescales of evolution and the illusion of precision. *Trends in Genetics*, 20, 80–86.

Hall, R. (1996). Reconstructing Cenozoic Southeast Asia. In *Tectonic Evolution of Southeast Asia*. Geological Society Special Publication no. 106 (ed. by Hall, R. & D. Blundell), pp. 153–184. The Geological Society, London.

Hall, R. (2002). Cenozoic geological and plate tectonic evolution of SE Asia and the SW Pacific: computer-based reconstructions and animations. *Journal of Asian Earth Sciences*, 20, 353–434.

Heads, M. (2005a). Dating nodes on molecular phylogenies: a critique of molecular biogeography. *Cladistics*, 21, 62–78.

Heads, M. (2005b). Towards a panbiogeography of the seas. *Biological Journal of the Linnean Society*, 84, 675–723.

Hennig, W. (1966). *Phylogenetic Systematics*. University of Illinois Press, Urbana, Illinois.

Holder, M.T., Erdmann, M.V., Wilcox, T.P., Caldwell, R.L. & Hillis, D.M. (1999). Two living species of coelacanths? *Proceedings of the National Academy of Sciences USA*, 96, 12616–12620.

Hovenkamp, P. (1997). Vicariance events, not areas, should be used in biogeographical analysis. *Cladistics*, 13, 67–79.

Hubbs, C.L. & Wisner, R.L. (1980). Revision of the sauries (Pisces, Scomberesocidae) with descriptions of two new genera and one new species. *Fisheries Bulletin*, 77,521–566.

Humphries, C.J. (2004). From dispersal to geographic congruence: comments on cladistic biogeography in the twentieth century. *Milestones in Systematics*. Systematics Association special volume 67 (ed. by Williams, D.M. & Forey, P.L.), pp. 225–260. CRC Press, Boca Raton, FL.

Humphries, C.J. & Parenti, L.R. (1986). *Cladistic Biogeography*. Oxford monographs on biogeography no. 2. Clarendon Press, Oxford.

Humphries, C.J. & Parenti, L.R. (1999). *Cladistic Biogeography, Interpreting Patterns of Plant and Animal Distributions*, 2nd ed., Oxford University Press, Oxford.

Inoue, J.G., Miya, M., Venkatesh, B. & Nishida, M. (2005). The mitochondrial genome of Indonesian coelacanth *Latimeria menadoensis* (Sarcopterygii: Coelacanthiformes) and divergence time estimation between the two coelacanths. *Gene,* 349, 227–235.

Kluge, A.G. (1988). Parsimony in vicariance biogeography: a quantitative method and a Greater Antillean example. *Systematic Zoology,* 37, 315–328.

Lieberman, B.S. (2000). *Paleobiogeography: Using Fossils to Study Global Change, Plate Tectonics, and Evolution* (Topics in Geobiology; v. 16). Kluwer Academic/Plenum Publishers, New York.

Lieberman, B.S. (2004). Range expansion, extinction, and biogeographic congruence: a deep time perspective. *Frontiers of Biogeography. New Directions in the Geography of Nature* (ed. by Lomolino, M.V. & Heaney, L.R.), pp. 111–124. Sinauer Associates, Inc., Sunderland, MA.

Lieberman, B.S. & Eldredge, N. (1996). Trilobite biogeography in the Middle Devonian: geological processes and analytical methods. *Paleobiology,* 22, 66–79.

Lomolino, M.V., Sax, D.F. & Brown, J.H. (2004). *Foundations of Biogeography, Classic Papers with Commentaries.* University of Chicago Press, Chicago, IL.

Lovejoy, N.R. & Collette, B.B. (2001). Phylogenetic relationships of New World needlefishe (Teleostei: Belonidae) and the biogeography of transitions between marine and freshwater habitats. *Copeia,* 2001, 324–338.

Lundberg, J.G. (1993). African-South American freshwater fish clades and continental drift: problems with a paradigm. In *Biological Relationships Between Africa and South America* (ed. by Goldblatt, P.), pp. 156–199. Yale University Press, New Haven, CT.

McCarthy, D. (2003). The trans-Pacific zipper effect: disjunct sister taxa and matching geological outlines that link the Pacific margins. *Journal of Biogeography,* 30, 1545–1561.

McCarthy, D. (2005). Biogeographical and geological evidence for a smaller, completely-enclosed Pacific Basin in the Late Cretaceous. *Journal of Biogeography,* 32, 2161–2177.

McDowall, R.M. (1997). The evolution of diadromy in fishes (revisited) and its place in phylogenetic analysis. *Reviews in Fish Biology and Fisheries,* 7, 443–462.

McLennan, D.A. (1994). A phylogenetic approach to the evolution of fish behaviour. *Reviews in Fish Biology and Fisheries,* 4, 430–460.

McLennan, D.A & Brooks, D.R. (2002). Complex histories of speciation and dispersal in communities: a re-analysis of some Australian bird data using BPA. *Journal of Biogeography,* 29, 1055–1066.

Meyer, C.P., Geller, J.B. & Paulay, G. (2005). Fine scale endemism on coral reefs: archipelagic differentiation in turbinid gastropods. *Evolution,* 59, 113–125.

Mooi, R.D. & Gill, A.C. (2002). Historical biogeography of fishes. In *Handbook of Fish and Fisheries Biology, Vol. 1 Fish Biology* (ed. by Hart, P.J.B. & Reynolds, J.D.), pp. 43–68. Blackwell Publishing, Oxford.

Morrone, J.J. (2002). Biogeographical regions under track and cladistic scrutiny. *Journal of Biogeography,* 29, 149–152.

Morrone, J.J. & Carpenter, J.M. (1994). In search of a method for cladistic biogeography: an empirical comparison of Component Analysis, Brooks Parsimony Analysis, and Three-area statements. *Cladistics,* 10, 99–153.

Murray, A. (1866). *The Geographical Distribution of Mammals.* Day and Son, Ltd, London.

Nelson, G. (1969). The problem of historical biogeography. *Systematic Zoology,* 18, 243–246.

Nelson, G. (1974). Historical Biogeography: an alternative formalization. *Systematic Zoology,* 23, 555–558.

Nelson, G. (1978). From Candolle to Croizat: comments on the history of biogeography. *Journal of the History of Biology,* 11, 269–305.

Nelson, G. (2004). Cladistics: its arrested development. In *Milestones in Systematics.* Systematics Association special Volume 67 (ed. by Williams, D.M. & Forey, P.L.), pp. 127–147. CRC Press, Boca Raton, FL.

Nelson, G., Atz, J.W., Kallman, K.D. & Smith, C.L. (1987). Donn Eric Rosen 1929–1986. *Copeia,* 1987, 541–547.

Nelson, G. & Ladiges, P.Y. (1991). Three-area statements: standard assumptions for biogeographic analysis. *Systematic Zoology,* 40, 470–485.

Nelson, G. & Ladiges, P.Y. (1993). Missing data and three-item analysis. *Cladistics,* 9, 111–113.

Nelson, G. & Ladiges, P.Y. (1996). Paralogy in cladistic biogeography and analysis of paralogy-free subtrees, *American Museum Novitates,* 3167, 1–58.

Nelson, G. & Ladiges, P.Y. (2001). Gondwana, vicariance biogeography and the New York school revisited. *Australian Journal of Botany,* 49, 389–409.

Nelson, G. & Platnick, N.I. (1981). *Systematics and Biogeography: Cladistics and Vicariance.* Columbia University Press, New York.

Nelson, G. & Rosen, D.E. (1981). *Vicariance Biogeography, a Critique.* Columbia University Press, New York.

Page, R.D.M. (1989). COMPONENT Version 1.5. University of Auckland, Auckland.

Page, R.D.M. (1990). Component analysis: a valiant failure? *Cladistics,* 6, 119–136.

Page, R.D.M. (1993). COMPONENT, Version 2.0. The Natural History Museum, London.

Parenti, L.R. (1981). [Discussion of] Methods of paleobiogeography. In *Vicariance Biogeography: A Critique* (ed. by Nelson, G. & Rosen, D.E.), pp. 490–497. Columbia University Press, New York.

Parenti, L.R. (1991). Ocean basins and the biogeography of freshwater fishes. *Australian Systematic Botany,* 4, 137–149.

Parenti, L.R. (in press). Life history patterns and biogeography: an interpretation of diadromy in fishes. *Annals of the Missouri Botanical Garden,* in press.

Parenti, L.R. & Humphries, C.J. (2004). Historical biogeography, the natural science. *Taxon,* 53, 899–503.

Patterson, C. (1975). The distribution of Mesozoic freshwater fishes. *Memoires du Museum National d'histoire Naturelle. Nouvelle Serie. Serie A. Zoologie,* 88, 156–174.

Patterson, C. (1981). Methods of paleobiogeography. In *Vicariance Biogeography: A Critique* (ed. by Nelson, G. & Rosen, D.E.), pp. 446–489. Columbia University Press, New York.

Platnick, N.I. & Nelson, G. (1978). A method of analysis for historical biogeography. *Systematic Zoology,* 27, 1–16.

Pouyaud, L., Wirjoatmodjo, S., Rachmatika, I., Tjakrawidjaja, A., Hadiaty, R. & Hadi, W. (1999). Une nouvelle espèce de coelacanthe. Preuves génétiques et morphologiques. *Comptes rendus de l'Académie des sciences. Serie III-Sciences de la Vie,* 322, 261–267.

Rosen, D.E. (1974). Phylogeny and zoogeography of salmoniform fishes and the relationships of *Lepidogalaxias salamandroides. Bulletin of the American Museum of Natural History,* 153, 265–326.

Rosen, D.E. (1976). A vicariance model of Caribbean biogeography. *Systematic Zoology,* 24, 431–464.

Rosen, D.E. (1978). Vicariant patterns and historical explanation in biogeography. *Systematic Zoology,* 27, 159–188.

Rosen, D.E. (1979). Fishes from the uplands and intermontane basins of Guatemala: revisionary studies and comparative geography. *Bulletin of the American Museum of Natural History,* 162, 267–376.

Rosen, D.E. (1985). Geological hierarchies and biogeographic congruence in the Caribbean. *Annals of the Missouri Botanical Garden,* 72, 636–659.

Rotondo, G.M., Springer, V.G., Scott, G.A.J. & Schlanger, S.O. (1981). Plate movement and island integration — a possible mechanism in the formation of endemic biotas, with special reference to the Hawaiian Islands. *Systematic Zoology,* 30, 12–21.

Sanmartín, I. & Ronquist, F. (2004). Southern hemisphere biogeography inferred by event-based models: plant versus animal patterns. *Systematic Zoology,* 53, 216–243.

Sclater, P.L. (1858). On the general geographical distribution of the members of the class Aves. *Journal of the Linnean Society of London (Zoology),* 2, 130–145.

Sparks, J.S. & Smith, W.L. (2005). Freshwater fishes, dispersal ability, and nonevidence: "Gondwana Life Rafts" to the rescue. *Systematic Biology,* 54, 158–165.

Springer, V.G. (1982). Pacific plate biogeography, with special reference to shorefishes *Smithsonian Contributions to Zoology,* 367, 1–182.

Springer, V.G. (1999). Are the Indonesian and western Indian Ocean coelacanths conspecific a prediction. *Environmental Biology of Fishes,* 54, 453–456.

Stearley, R.L. (1992). Historical ecology of Salmoninae, with special reference to Oncorhynchus. In *Systematics, Historical Ecology and North American Freshwater Fishes* (ed. by Mayden, R.L.), pp. 622–658. Stanford University Press, Stanford, CA.

Thacker, C.E. & Hardman, M.A. (2005). Molecular phylogeny of basal gobioid fishes Rhyacichthyidae, Odontobutidae, Xenisthmidae, Eleotridae (Teleostei: Perciformes: Gobioidei). *Molecular Biology and Evolution,* 37, 858–871.

Upchurch, P., Hunn, C.A. & Norman, D.B. (2002). An analysis of dinosaurian biogeography: evidence for the existence of vicariance and dispersal patterns caused by geological events. *Proceedings of the Royal Society of London B Biological Sciences,* 269, 613–621.

Van Veller, M.G.P. & Brooks, D.R. (2001). When simplicity is not parsimonious: a priori and a posteriori methods in historical biogeography. *Journal of Biogeography,* 28, 1–11.

Van Veller, M.G.P., Korner, D.J. & Zandee, M. (2000). Methods in vicariance biogeography: assessment of the implementation of assumptions zero, 1 and 2. *Cladistics,* 16, 319–345.

Vari, R.P. & Weitzman, S.H. (1990). A review of the phylogenetic biogeography of the freshwater fishes of South America. In *Vertebrates in the Tropics* (ed. by Peters, G.R. & Hutterer, R.), pp. 381–393. Museum Alexander Koenig, Bonn.

Voelker, G. (1999). Dispersal, vicariance, and clocks: historical biogeography and speciation in a cosmopolitan passerine genus (Anthus: Motacillidae), *Evolution,* 53, 1536–1552.

Wallace, A.R., (1863). On the physical geography of the Malay Archipelago. *Journal of the Royal Geographical Society,* 33, 217–234.

Wallace, A.R. (1876). *The Geographical Distribution of Animals* (2 vols.), Macmillan, London.

Wanntorp, L. & Wanntorp, H.-E. (2003). The biogeography of *Gunnera L.*: vicariance and dispersal. *Journal of Biogeography,* 30, 979–987.

Webster, B. (1979). New theory disputes old in geographical distribution of species. *The New York Times,* Tuesday, May 8, C2.

Wiley, E.O. (1987). Methods in vicariance biogeography. In *Systematics and Evolution: A Matter of Diversity* (ed. by Hovenkamp, P.), pp. 283–306. University of Utrecht, Utrecht.

Wiley, E.O. (1988). Parsimony analysis and vicariance biogeography, *Systematic Zoology,* 37, 271–290.

Winterbottom, R. & McLennan, D.A. (1993). Cladogram versatility: evolution and biogeography of acanthuroid fishes, *Evolution,* 47, 1557–1571.

3 A Brief Look at Pacific Biogeography: The Trans-Oceanic Travels of *Microseris* (Angiosperms: Asteraceae)

John R. Grehan

ABSTRACT

The modern revolution in biogeography did not begin with plate tectonics. It began two decades earlier when Leon Croizat established geographic distributions as the empirical foundation of evolutionary biogeography. Comparative map analysis reveals patterns that are not accessible through other methods. The biogeography of *Microseris* (Angiosperms: Asteraceae) is used to illustrate the power of geographic analysis to provide unique insights into the biogeographic distributions and relationships of organisms. Explanations of dispersal as physical movement for *Microseris* are shown to be problematic by the congruent distributions and Pacific homology of this genus with groups of diverse means of dispersal such as daisies, dragonflies, millipedes, eyebrights, and seaweeds. The role of tectonics and the historical implications of *Microseris* biogeography for molecular clock theory are briefly discussed.

INTRODUCTION

Biogeography has not changed much from the time of Darwin (1859). After 150 years of evolutionary biogeography, most biogeographers still follow Darwin's theory of geographic distribution requiring dispersal from geographically restricted centers of origin. Never mind whether the dispersal takes place before (process of vicariance) or after (process of ecological dispersal) the formation of barriers, the underlying principle of Darwinian biogeography is that organisms move about so their actual distributions are historically uninformative. To discern the geographic context of evolution, Darwinian biogeographers look to historical theories of ecology, systematics, molecular clocks, and geology — anything but distributions themselves — as the empirical data of biogeography.

The historical narratives of these non-biogeographic disciplines are transformed into actual reality as the platform upon which the biogeography plays itself out. And perhaps this is not surprising, as most of those who enter the field of biogeography are actually more interested in other aspects of evolution such as a particular group of organisms or a biological field such as ecology, systematics, or genetics. Interest in biogeography as such — the analysis of geographic relationships for life in general — is something that often does not enter into consideration. One only has to see how often distribution maps are left out of the picture, even for papers focusing on biogeographic theory and method.

MOLECULAR MYTHOLOGY

Over the last ten years or so, Darwinian biogeography has claimed a newfound sense of authority through the application of molecular clock theory as the final solution to that holy grail of Darwinian biogeography — the individual separation of dispersal and vicariance. Amidst a fanfare of rhetoric and propaganda these practitioners claim to have transformed biogeography (DeQueiroz, 2005; Didham, 2005; Waters, 2005), while in reality their approach is no more than a molecular application of Darwinian biogeography (Heads, 2005a). Molecular clocks are supposed to finally test the dispersal vs. vicariance problem of Darwinian biogeography according to whether the divergence estimates match or postdate a hypothesized geological event (Waters, 2005). The method is so simple that one can support its authority even without being a biogeographer (e.g., Didham, 2005), or criticize Croizat (1964) without reading the work — as did Waters & Roy (2004) based on a characterization of panbiogeography by Humphries (2000), an error acknowledged by J. Waters (pers. comm.).

Apart from the assumption that a geological theory has some kind of necessary empirical reality, the Darwinian practitioners usually overlook the fact that molecular clock scenarios based on minimum age of fossilization cannot generate maximum estimates of divergence. Even though molecular theorists admit as much, they somehow forget this maxim when final judgments are rendered upon group after group — with a common refrain being that the estimated divergence postdates theorized geological events and therefore falsify vicariance (Heads, 2005a).

GEOLOGY FIRST?

Biogeography does get complicated if one attempts to apply Darwinian theory to systematics and geography. The possibilities become almost as numerous as the number of branches of a cladogram. All of these approaches share in common the presumption that biogeographic reconstruction is confined to background knowledge derived from ecology, fossils, and stratigraphy all mixed together with various, and sometimes contradictory, notions of centers of origin and dispersal (Croizat, 1952). Islands such as the Galapagos are seen as self-evident proofs of long-distance dispersal followed by isolation and differentiation because they are obviously volcanic, oceanic in origin, and without any historical connections with continental mainland areas. Mayr (1982) made this argument when dismissing Croizat's (1958) support for ancient Galapagos plant and animal life being inherited as a whole from the American continent as "... so totally refuted by the geological and biological evidence that no further comment is necessary". Clearly the message here is that background knowledge is beyond challenge.

But even as Mayr wrote those words, geologists had already discovered that the Galapagos was an integral part of a major tectonic structure (the Galapagos Gore) extending between the East Pacific Rise and the Americas. This discovery corroborated Croizat's (1958) earlier prediction, made before seafloor mapping, that the Galapagos were associated with a major undiscovered tectonic center. This prediction was necessary because the biogeographic patterns involving the Galapagos conformed to a 'continental' pattern as if the Galapagos were embedded within a continent rather than the ocean. A number of geologists were also contemplating complex tectonic models that predicted Mesozoic and Cenozoic island arcs or micro continents in the eastern Pacific that overlapped the Galapagos and are now embedded within continental America. Some geological models even included connections with Asia across what is now the Pacific basin (Grehan, 2001a). These geological discoveries show that, just because a biogeographic hypothesis conflicts with current background knowledge, it does not mean to say the biogeography is necessarily wrong (in contrast to the belief of Mayr 1982).

A PRIMER IN BIOGEOGRAPHY

Above all, it is the lack of biogeographic knowledge that haunts biogeography. Ask the basic question, as did Croizat (1952) over half a century ago, whether certain place names such as Kerguelen, Madagascar, Mascarenes, Rapa, etc., bring forth a geographic image in the mind. Croizat (1952) saw the ability to answer such a seemingly simplistic question as the necessary starting point for following a discussion on dispersal. It was this fundamental shift that revolutionized biogeography, not the later acceptance of plate tectonics as belatedly claimed by Darwinian biogeographers. The revolution took place when, for the first time, **biogeographic maps** became the foundation of biogeographic analysis. Biogeographic maps are not biogeographic distributions; they are maps of the spatial structure and homology of distributions (Croizat, 1952; Craw et al., 1999; Grehan, 2001b). Distributions only document spatial location. Regrettably, many practitioners of biogeography seem to be little better prepared in biogeography now than in Croizat's time. Biogeographers often seem profoundly ignorant of the principal patterns of biogeography — the tracks and nodes of life that are the empirical traces of evolution in space/time — when they continue to dismiss individual distributions as puzzling or mysterious (Craw et al., 1999).

It is knowledge of distributions that may be the most powerful tool for any evolutionary biologist: "If species evolve from preexisting species, then the geographic distribution of species should tell us something about the actual course that phylogeny has taken" (Hull, 1988). Attention to actual distribution assumes that the geographic structure of distributions — the distributional limits, the spatial geometry, the boundaries and range, taxa present, taxa absent, etc., are all potentially informative about the evolutionary structure of life (Craw et al., 1999). If dispersal forever repeats (Croizat, 1964), then so too does biogeography. In the following section the consequences and the possibilities are considered for an obscure group of daisies that are supposed to be the result of the fortunes of ecology, but whose distributions may speak for a very different evolutionary history.

LOGIC OF DISPERSAL

Dispersal in evolutionary biogeography is all about the evolution of distributions or dispersal in the broad sense that comprises two empirical elements: relative locations and geographically differentiated biological forms ranging from individual genetic and morphological differences to those that are delineated by formal taxonomic ranks. Dispersal must, therefore, account for both spatial and biological differences. Darwinian biogeography attempted to encompass dispersal within the ecological meaning of the term to refer to physical movement through the active or passive migration of organisms. In this Darwinian context, dispersal is a concept borrowed from the science of ecology and refers to the movement of organisms.

Croizat (1964) recognized a biogeographic concept of dispersal that encompassed both the spatial and the biological differentiation of form, and he therefore proposed a purely biogeographic definition of dispersal as translation in space + form-making. Dispersal in this context is a biogeographic concept referring to the evolution of both spatial structure and biological differentiation of form (Grehan, 2001b). Evolution can, by itself, bring about the presence of a taxon at a particular location without physical movement. This is a radical departure from earlier convention, and perhaps Croizat could be criticized for not developing an entirely novel terminology. But in this logic of dispersal one can look at biogeographic evolution in terms of either mobility (ecological dispersal) or differentiation (biogeographic dispersal) or as a combination of the two. The following sections examine the different contexts of dispersal from Darwinian and panbiogeographic perspectives.

DISPERSAL THROUGH MIGRATION

Comprising about 13 species of annual and perennial daisies, the genus *Microseris* is generally found in grasslands or other open habitats from wooded to desert and scrub communities (Chambers, 1955). Three species are found in western North America. One species occurs in southern South America, another in New Zealand and Australia, and a third in eastern and Western Australia. In a paper presenting DNA evidence for *Microseris* evolving in Australia and New Zealand after long-distance dispersal from western North America, Vijverberg *et al.* (1999) note that adaptive radiations on various islands are supposed to have evolved "from one or few individuals after long distance dispersal", and that molecular diversificatio among oceanic island relatives suggests the sampling effect of genetic drift following a founding event. Having established these suppositions, they acknowledge Carlquist's (1983) documentation of intercontinental disjunctions between the west coast of North America and southern South America. Vijverberg *et al.* (1999) suppose these disjunctions are, like floras of the oceanic islands, the result of bird dispersal rather than tectonic plate movement.

Here is the reasoning. First, suppose adaptive radiations on oceanic islands are the result of chance long-distance dispersal. Then, suppose a low level of molecular diversification is evidence of long-distance dispersal. And also suppose that long-distance disjunctions between western North and South America are the result of the same process. One has effectively already arrived at the conclusion of long-distance dispersal whereby the DNA result will, by definition, provide evidence of that dispersal. Vijverberg *et al.* (1999) illustrate this reasoning in characterizing the Australian and New Zealand *Microseris* species as a pattern of adaptive radiation following intercontinental dispersal of an annual and perennial North American hybrid that reached Australia and New Zealand by long distance dispersal. After constructing a chloroplast DNA phylogeny showing the Australian and New Zealand species to be a monophyletic clade, Vijverberg *et al.* (1999) present this phylogeny as evidence for the distribution originating from a single or a few closely spaced colonizing events. The single origin was anticipated because of the supposed hybrid origin in North America, and the uniformity of its allotetraploid karyotype was seen to be compatible with this process.

The study by Vijverberg *et al.* (1999) repeats similar contentions made by Bachman (1983), who also accepted the migrations of *Microseris* as a biogeographic fact. He reports on two "unrelated" long-distance dispersal events, one for Australia and New Zealand, and one for Chile. The latter was considered to be "less spectacular and easier to analyze". Having accepted long-distance dispersal, the developmental genetics of *Microseris* was interpreted both as the result of this process as well as "testing" the hypothesis itself. Here a theoretical conjecture is transformed into fact informing empirical studies of genetics that in turn are supposed to give insight into the process theorized in the first place. There is the very real danger of making a round trip back and forth from one theory to another.

Carlquist (1983) argued for intercontinental dispersal between western North and South America, because separation of tectonic plates could not explain the disjunction. He also saw the ability to hybridize species from the two continents and the occurrence of conspecifics as evidence that these migrations occurred recently. Migratory birds were the obvious link, and this mode of theorized migration was congruent with a seed and fruit morphology that would allow transport through ingestion or external attachment.

MICROSERIS (PANBIO)GEOGRAPHY

According to Chambers (1963), the single southern South American species is the result of bird dispersal carrying seeds from North America on a one-way trip that happened to occur in a variety of plant groups. This might be possible, but what about the other side of the Pacific? The species *M. lanceolata* is found only in Australia, and then only above 200 m in certain parts of Australia: southwestern Australia above 200 m at Esperance, and southeastern Australia from Spencer Gulf inland southeast to the east and west of Melbourne, and northeast to Bundarra and Armidale in New South Wales. The distribution vicariates with another New South Wales species (*M. scapigera*) centered on localities around Melbourne, central Tasmania, and New Zealand east of the Southern Alps in the South Island near Mt.

Cook to the Kaikouras, and in the North Island from the southeastern coast to inland Tongariro. Again, what is the biogeographic meaning of this pattern? What is the biogeographic meaning of these localities? How is it that the species come to be just where they are?

The distribution of *Microseris* within North America may not seem all that significant for biogeography in general. After all, there are many other plants and animals with similar distributions. But that is just the point. Why does the genus have a North American range limited to the western region? If biogeography is the combined act of individual and unrelated histories, such questions may not have general significance After all, in looking to the migratory abilities of animals and plants anything is possible, and distributional limits are simply the result of chance. This would leave *Microseris* within an evolutionary universe of its own, and its biogeographic history is that of itself. No other organisms need matter unless they are providing the means of dispersal such as birds carrying *Microseris* seeds. And yet, to the biogeographer these localities may be as telling about *Microseris* as they are about the evolution of distributions in general because they are not biogeographic features of *Microseris* alone. Chambers (1955) hinted at this possibility when he predicted that a better knowledge of the origin of *M. scapigera* may have very interesting consequences for studies of plant distribution and relationships in the southern hemisphere.

What is interesting to the (pan)biogeographer is the geographic pattern. *Microseris* has a classic biogeographic range with respect to all its main locations. Its main massing (geographic concentration of diversity) is unambiguously around the Pacifi basin (Figure 3.1). All species are located adjacent to, or closer to the Pacific basin than any other ocean basin. This is a spatial homology. It cannot be derived from any method that does not incorporate spatial characters. It cannot be derived from an area cladogram constructed from biological relationships. With a spatial homology one may recognize not only the shared biogeographic homology of *Microseris* with other Pacific distributions, but also individual localities. The pattern of biogeographic dispersal (as translation in space + form-making) involving Western North America, southern South America (in this case west of the Andes) and Australia/New Zealand is biogeographically standard (Craw et al., 1999). The pattern is found in organisms with such diverse means of dispersal as petalurid dragonflies, heterochordeumatoid millipedes, eyebrights (*Euphrasia*), and blennospermatinae daisies (Figure 3.2). All of these distributions are more widespread than that of *Microseris,* but they share the localized ranges of Western North America, southern South America, and Australia. In detail, the distributions of Heterochordeumatoidea, *Euphrasia,* and *Abrotanella* share with *Microseris* the southern Australian range of Western Australia and the connection between Tasmania and New South Wales. *Blennosperma* and *Microseris* each occur in both Western North America and southern South America, but *Blennosperma* also vicariates with *Crocidium* and *Abrotanella* in North and South America, respectively (Heads, 1999). For all that one might theorize about *Microseris* being ferried by birds, the Pacific biogeography of *Microseris* is biogeographically as one with millipedes, dragonflies, eyebrights, and blennospermatid daisies.

Perhaps the coincidence of distributions is purely chance. As one contributor recently commented on TAXACOM, even blind pigs can occasionally find acorns. Well, perhaps. Whether 'coincidence' or not, a biogeography of *Microseris* must

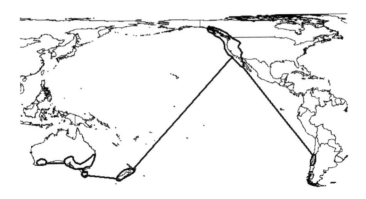

FIGURE 3.1 Pacific dispersal (translation in space) of *Microseris,* showing main massing of distribution around the Pacific basin with local massing in Western North America (about 13 species), South West Pacific (about two species), and southern South West South America (one species which is also recorded near Lima, Peru. This location is not included on the map at this time, pending further corroboration of its evolutionary status — see Lohwasser *et al.* (2004). Tracks drawn across the Pacific reflect published opinion about affinities between species at each location.

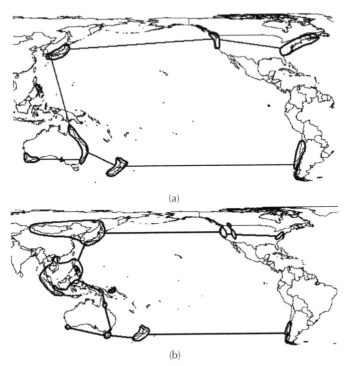

(a)

(b)

FIGURE 3.2 Pacific dispersal (translation in space) of: (a) petalurid dragonflies (Eskov and Golovatch 1986; Watson et al., 1991; Carle, 1995; Davies 1998), (b) heterochordeumatoid millipedes (Shear, 1999), (c) eyebrights (*Euphrasia*) (Barker, 1982), and (d) Blenospermatinae daisies (Heads, 1999).

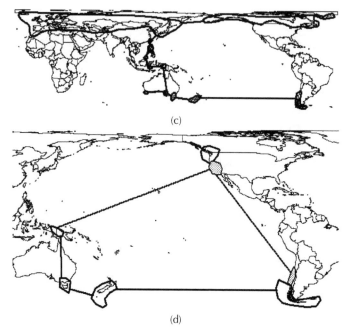

FIGURE 3.2   (Continued)

also be satisfactory for millipedes, dragonflies, eyebrights, and everything else. One might ask, in this respect, whether it is just meaningless coincidence that the North and South American connection is also found in marine organisms such as the seaweed *Macrosytus,* which is also found in southern New Zealand and Tasmania as well as other islands in the south Atlantic and Indian Ocean (Chin et al., 1991). An appeal to bird dispersal for *Microseris* is not very satisfactory in these respects. It might 'explain' the disjunctions if one could imagine eyebrights, millipedes, dragonflies, and seaweeds taking the same trip, but not the geological correlations.

And what about the geology? Plate tectonics were ruled out for *Microseris* by Carlquist (1983) and Vijverberg *et al.* (1999), but plate tectonics and *Microseris* do share some geographic features. The region of Western North America was predicted by Croizat (1958) to comprise a tectonically distinct area with strong Pacific relationships. This prediction was later corroborated by geologists who recognized that the region comprises tectonic terranes of Pacific origin (Craw et al., 1999). In addition, southern South America, New Zealand, and southeastern Australia all comprise terranes with Pacific affinities (Craw et al., 1999; Heads, 1999). The northern limits of *Microseris* in Australia happen to coincide with one of the continent's pre-eminent biogeographic landmarks — the MacPherson-Macleay Overlap (Ladiges, 1998). This biogeographic node locates the intersection of biogeographic tracks which run throughout the region, and across the Pacific (Croizat 1964), and it is also characterized by accreted terranes from the Tasman orogen (Heads, 2001).

The local and global spatial correlations between the distribution of *Microseris* and tectonic structure suggest the possibility of a shared geological and biological history as a key to its geographic evolution both around the Pacific and between North and South America. The role of tectonic plate movements cannot, therefore, be ruled out as suggested by Carlquist (1983). In an evolutionary sense the biogeography of the present represents the ecology of the past (Craw et al., 1999). The predictive success (blind pigs aside) for correlating distributions with tectonic history in panbiogeography does show that biogeographic theory was ahead of the background geological knowledge for that time (Craw and Weston, 1984; Craw and Page, 1988). The method works.

DISPERSAL THROUGH FORM-MAKING

An evolutionary model of dispersal that would account for *Microseris* as well the other Pacific distributions is vicariant form-making. *Microseris* species occur where they do because the immediate ancestor of all *Microseris* occurred in all these localities together. *Microseris* is distributed around the Pacific because the main massing of its common ancestor was distributed over the Pacific (this does not mean that the entire range of this ancestor, or that of related genera was necessarily Pacific), and each species originated in its current region without individually dispersing between them. In Australia and New Zealand *Microseris* still retains the ancestral vicariism even though one species has become disjunct following the separation of New Zealand and Australia by the Tasman Sea at the close of the Mesozoic. This pattern of vicarious descent within the ancestral main massing is a general biogeographic process repeated in other broadly distributed groups such as the *Nothofagus* alliance (Heads, 2006).

The Pacific disjunction of *Microseris* reflects loss of former geographic continuity that may be due to any number of geological mechanisms such as mobile continents, micro-continents, or island arcs foundering, or land-bridges through an expanding Earth. Whatever the geological model, it is the tectonic correlation that is historically predictive. It is quite possible that the ancestor of *Microseris* was more widely distributed and occurred in other areas around the Pacific basin. The panbiogeographic model predicts that the ancestor of *Microseris* was less likely to have occurred in central Gondwana regions of Africa, India, and eastern America, while more likely to have occurred in eastern Asia. The correlation suggests the evolutionary origin of *Microseris* may be as old as the Pacific basin — the Jurassic — or that it represents a later period of mobilism correlated with subsequent tectonic history in the Pacifi region in the Mesozoic. Molecular clock estimates are currently unresolved for *Microseris* (Wallace & Jansen 1999), but post-Mesozoic divergence estimates would not falsify the tectonic prediction unless an earlier molecular divergence is predicted — assuming that the molecular clock is constant (cf. Heads, 2005a).

PAST, PRESENT, FUTURE

Biogeographers need not reduce themselves to being the inferior subordinates of geologists, molecular theorists, or historical ecologists when it comes to generating new knowledge. Progress in science always comes from uncertainty at the edge (Craw & Heads, 1988). At the center of scientific popularity one may find life to be quiet and predictable, and also rather dull. I first looked at the biogeographic edge of biology as a graduate student when I became aware from Robin Craw and Michael Heads that there were unanswered questions in evolution, and simply ignoring these questions (which was the standard approach of the time) was not a realistic scientifi solution. I found it was not necessary to follow the accepted views on evolution simply because they were stated by prominent and influential evolutionists. I could follow a path in which biogeographic maps represented the real scientific authority.

But there is a considerable price for contesting established ideas (Colacino & Grehan, 2003). Croizat paid that price in losing his position at the Arnold Arboretum for the simple crime of publishing a critique on botanical theory that was supported by I. W. Bailey (Craw, 1984). Biogeographic exploration has since been stultifie in New Zealand (Heads, 2005b). With publication in journals such as *Candollea, Cladistics,* the botanical and biological journals of the Linnean Society, *Journal of Biogeography,* and *Telopea,* the situation has improved outside New Zealand since the early 1980s when *Systematic Zoology* was the only international English language journal that would consider discussing panbiogeography. Emergence of the Latin American 'school' of panbiogeography has also generated many publications (see, e.g., Morrone & Llorente, 2003; Llorente & Morrone, 2005). However, there are new challenges ahead, as some molecular biologists and editors believe that molecular clocks have now falsified both the panbiogeographic method and the panbiogeographic synthesis.

ACKNOWLEDGEMENTS

I am grateful to Kenton Chambers for providing factual corroboration on *Microseris* taxonomy and distribution, some very informative feedback on the biogeographic perspective presented here, and for raising broader questions that could not be addressed here. I am also indebted to Dennis McCarthy, Michael Heads, and J. Bastow Wilson for helpful review and valuable insights.

REFERENCES

Bachman, K. (1983). Genetic analysis of evolution associated with dispersal in plants. *Sonderbäde des Naturwissenschaftlichen Vereins in Hamburg,* 7, 65–86.

Barker, W.R. (1982). Taxonomic studies in *Euphrasia* L. (Scrophulariaceae). A revised infrageneric classification, and a revision of the genus in Australia. *Journal of the Adelaide Botanical Garden,* 5, 1–304.

Carle, F.L. (1995). Evolution, taxonomy, and biogeography of ancient gondwanian libelluloides, with comments on anisopteroid evolution and phylogenetic systematics (Anisoptera: Libelluloidea). *Odonatologica,* 14, 383–424.

Carlquist, S. (1983). Intercontinental dispersal. *Sonderbäde des Naturwissenschaftlichen Vereins in Hamburg,* 7, 37–47.

Chambers, K.L. (1955). A biosystematic study of the annual species of *Microseris. Contributions from the Dudley Herbarium,* 4, 207–312.

Chambers, K.L. (1963). Amphitropical species pairs in *Microseris* and *Agoseris* (Compositae: Cichorieae), *The Quarterly Review of Biology,* 38, 124–140.

Chin, N.K.M., Brown, M.T. & Heads, M.J. (1991). The biogeography of Lessoniaceae, with special reference to *Macrocystis* C. Agardh (Phaeophyta: Laminariales). *Hydrobiologia*, 215, 1–11.

Colacino, C. & Grehan, J.R., (2003). Ostracismo alle frontiere della biologia evoluzionistica: il caso Léon Croizat. In *Scienza e Democrazia* (ed. by Capria, M.M.), pp. 195–220. Ligouri Editore, Napoli.

Craw, R.C. (1984). Never a serious scientist: the life of Leon Croizat. *Tuatara*, 27, 5–7.

Craw, R. & Heads, M.J. (1988). Reading Croizat: on the edge of biology. *Rivista di Biologia Biology Forum*, 81, 499–532.

Craw, R.C. & Page, R. (1988). Panbiogeography: method and metaphor in the new biogeography. In *Evolutionary Processes and Metaphors* (ed. by Ho, M.-W. & Fox, P.), pp. 163–189. John Wiley & Sons, Chichester.

Craw, R.C. & Weston, P. (1984). Panbiogeography: progressive research program? *Systematic Zoology*, 33, 1–13.

Craw, R.C., Grehan, J.R. & Heads, M.J. (1999). *Panbiogeography: Tracking the History of Life*. Oxford University Press, New York.

Croizat, L. (1952). *Manual of Phytogeography*, Junk, The Hague.

Croizat, L. (1958). *Panbiogeography*, published by the author, Caracas.

Croizat, L. (1964). *Space, Time, Form, The Biological Synthesis*, published by the author, Caracas.

Darwin, C. (1859). *The Origin of Species*, John Murray, London.

Davies, D.A.L. (1998). The genus *Petalura:* field observations, habits and conservation status (Anisoptera: Petaluridae). *Odanatologica*, 27, 287–305.

DeQueiroz, A. (2005). The resurrection of oceanic dispersal in historical biogeography. *Trends in Ecology and Evolution*, 20, 68–73.

Didham, R.K. (2005). New Zealand: 'fly paper of the Pacific'? *The Weta*, 29, 1–5.

Eskov, K.Y. & Golovatch, S.I. (1986). On the origin of trans-Pacific disjunctions. *Zoologische Jahrbä her Systematik*, 118, 265–285.

Grehan, J.R. (2001a). Biogeography and evolution of the Galapagos: integration of the biological and geological evidence. *Biological Journal of the Linnean Society*, 74, 267–287.

Grehan, J.R. (2001b). Panbiogeography from tracks to ocean basins: evolving perspectives. *Journal of Biogeography*, 28, 413–429.

Heads, M.J. (1999). Vicariance biogeography and terrane tectonics in the South Pacific analysis of the genus *Abrotanella* (Compositae). *Biological Journal of the Linnean Society*, 67, 391–432.

Heads, M.J. (2001). Birds of paradise (Paradisaeidae) and bowerbirds (Ptilonorhynchidae): regional levels of biodiversity and terrane tectonics in New Guinea. *Journal of Zoology London*, 255, 331–339.

Heads, M.J. (2005a). Dating nodes on molecular phylogenies: a critique of molecular biogeography. *Cladistics*, 21, 62–78.

Heads, M.J. (2005b). The history and philosophy of panbiogeography. In *Regionalizació Biogeográfica en Iberoamérica y Tópic os Afines* (ed. by Llorente, J. & Morrone, J.J.), pp. 67–123. Universidad Nacional Autónoma de México, México.

Heads, M.J. (2006). Panbiogeography of *Nothofagus* (Nothofagaceae): analysis of the main species massings. *Journal of Biogeography*, 33, 1066–1075.

Hull, D. (1988). *Science as a Process*. The University of Chicago Press, Chicago and London.

Humphries, C.J. (2000). Form, space, and time: which comes first? *Journal of Biogeography* 27, 11–15.

Ladiges, P.Y. (1998). Biogeography after Burbidge. *Australian Systematic Botany*, 11, 231–242.

Llorente, J. & Morrone, J.J. (2005). *Regionalizació Biogeogrfic a en Iberoamérica y Tpi cos Afines,* Universidad Nacional Autónoma de México, México.

Lohwasser, U., Granda, A. & Blattner, F.R. (2004). *Phylogenetic analysis of Microseris (Asteraceae), including a newly discovered Andean population from Peru. Systematic Botany,* 29, 774–780.

Mayr, E. (1982). Vicariance biogeography [review]. *Auk,* 99, 618–620.

Morrone, J.J. & Llorente, J. (2003). *Una Perspectiva Latinoamerica de la Biogeografía.* Universidad Nacional Autónoma de México, México.

Shear, W.A. (1999). Millipeds. *American Scientist,* 87, 232–239.

Vijverberg, K., Mes, T.H.M., & Bachmann, K. (1999). Chloroplast DNA evidence for the evolution of *Microseris* (Asteraceae) in Australia and New Zealand after long-distance dispersal from western North America. *American Journal of Botany,* 86, 1448–1463.

Wallace, R.S. & Jansen, R.K. (1990). Systematic implications of chloroplast DNA variation in the genus Microseris (Asteraceae: Lactuaceae). *Systematic Botany,* 15, 606–616.

Waters, J.M. (2005). Historical biogeography: an introduction. *Systematic Biology,* 54, 338–240.

Waters, J.M., and Roy, M.S. (2004). Out of Africa: the slow train to Australasia. *Systematic Biology,* 53, 18–24.

Watson, J.A.L., Theischinger, G. & Abbey, H.M. (1991). *The Australian Dragonfl es.* CSIRO, Canberra & Melbourne.

4 Biotic Element Analysis and Vicariance Biogeography

Bernhard Hausdorf and Christian Hennig

ABSTRACT

Clustering of distribution areas is one of the most prominent biogeographic patterns. One mechanism that might cause this pattern is vicariance. According to the vicariance model, diversification is the result of a fragmentation of an ancestral biota by emerging barriers. Such a fragmentation will result in the formation of distinct biotic elements (i.e., groups of species with similar ranges) in areas of endemism. We use biotic elements as biogeographic units, because areas of endemism cannot be delimited if dispersal occurred. We propose a statistical test for clustering of distribution areas based on a Monte Carlo simulation with a null model that generates range data sets such that their range size distribution, the species richness distribution of the geographic cells, and the spatial autocorrelation of the occurrences of a taxon approach the parameters in the real data set. Biotic elements were delimited with model-based Gaussian clustering. We also tested another prediction of the vicariance model, namely that closely related species originate on different sides of an emerging barrier and, hence, belong to different biotic elements.

The ranges of the land snail species of the central Aegean Islands are not significantly clustered. The ranges of north–west European land snails and Iberian land snails belonging to the Helicoidea are significantly clustered, but contrary to the prediction of the vicariance model, closely related species belong significantly more often to the same biotic element than should be expected by chance. Only a data set including land snail species with restricted distributions in Israel and Palestine meets both tested predictions of the vicariance model: ranges are significantly clustered, and closely related species are distributed across biotic elements. These case studies indicate that speciation modes other than vicariance were frequent or that the imprint of vicariance on the ranges was often obscured by extensive postspeciational dispersal or regional extinction.

INTRODUCTION

The theme of the symposium where this paper has been presented was "What is Biogeography?" Thus, we start with a definition of biogeography, and we will show how biotic element analysis fits into that framework.

Biogeography is the discipline that investigates the spatial distribution of organisms and of attributes of organisms, for example, the variation of body size with latitude (e.g., Bergmann's rule) or the increase of latitudinal range with latitude (Rapoport's rule). It looks for spatial patterns and tries to elucidate the processes and mechanisms that cause such patterns.

Biotic element analysis is an approach to investigate patterns in the distribution of organisms themselves, namely a clustering of distribution areas. There are several processes that might result in this pattern. Clustering of distribution areas can be the result of dispersal, of regional extinction or of speciation. It might originate if an ecologically more or less homogeneous area is colonized by different species, even if these species originated in different regions and have different histories. It might originate if species with originally different distribution areas became restricted to the same area ("refuge"), for example, as a result of climatic detorioration. Or it might originate if the populations of several stem species present in an area evolve into new species. The most explicit model that postulates the origin of several species in the same region is the vicariance model.

The aim of our paper is to assess the importance of vicariance in shaping biotas. We derive predictions from the vicariance model that can be tested. We show that areas of endemism cannot be used as biogeographic units in tests of these predictions, because they cannot be delimited if dispersal occurred. We show that biotic elements are suitable units for tests of the vicariance model. We investigate the importance of vicariance in some case studies of land snail distributions.

THE VICARIANCE MODEL

According to the vicariance model (Croizat et al., 1974; Rosen, 1976, 1978; Platnick & Nelson, 1978; Nelson & Platnick, 1981; Wiley, 1988; Humphries & Parenti, 1999), an ancestral biota was fragmented by the appearance of a barrier. The barrier interrupted the gene flow between the populations separated by the barrier and, consequently, this vicariance event resulted in allopatric speciation of many of the species formerly constituting the ancestral biota. In this way, two new biotas originated, separated by the barrier. By repetitions of the described process, areas of endemism with distinct biotas (i.e., with many species restricted to the individual areas) emerged. On average, the ranges of the species which originated in the same area of endemism will be more similar to each other than to ranges of species which originated in other areas of endemism. Thus, the vicariance model predicts a clustering of species ranges (Morrone, 1994; Hausdorf, 2002). The species originating in an area of endemism by vicariance form a biotic element, a group of taxa which ranges are more similar to each other than to those of other such groups (Hausdorf, 2002).

The pattern for which we can examine distribution data is clustering of distribution areas. Areas of endemism as such cannot be found in distribution data.

They can only be inferred from clusters of species ranges. This distinction is obscured in the operational methods for identifying and delimiting areas of endemism (Morrone, 1994; Linder, 2001; Szumik et al., 2002; Mast & Nyffeler, 2003; Szumik & Goloboff, 2004).

The delimitation of areas of endemism is not problematic as long as species do not disperse across the barriers separating the areas of endemism. Under these conditions, the borders of the areas of endemism can be drawn between the range clusters. An example might be widely separated oceanic islands. However, in most cases, there is stochastic dispersal of species across barriers with time. Moreover, many barriers weaken or disappear with time. If dispersal across the barriers that separated the areas of endemism resulted in an overlap of ranges of species that originated in different areas of endemism, biogeographical data alone are insufficient for delimiting areas of endemism or biogeographic regions (Hausdorf, 2002). A well-known example, which highlights this problem, is the number of lines which have been proposed to delimit the Oriental and the Australian region (Mayr, 1944; Holloway & Jardine, 1968; Simpson, 1977; Vane-Wright, 1991). If it is not possible to establish the border between two regions that were separated by several hundred kilometers of sea on the basis of biogeographical data, it will hardly be possible to delimit areas of endemism that were separated only temporarily by climatic barriers on a continent, for example. It is possible to formulate optimality criteria to choose among different delimitations of areas (Szumik et al., 2002; Szumik & Goloboff, 2004). But the resulting borders vary with the choice of the criteria, and there is no rationale how to find the true borders. Therefore, areas of endemism can hardly be used as units in tests of the vicariance model.

On the contrary, biotic elements can be recognized, even if some of the species that originated by vicariance dispersed across the barriers that separated the areas of endemism. The existence of biotic elements is predicted by the vicariance model, and in contrast to areas of endemism, biotic elements can be determined by means of distribution data alone. Hence, they are suitable as units in tests of the vicariance model.

Usually vicariance cannot be observed directly, because this process generally takes geologic time periods. Therefore, it is important to derive predictions about observable patterns from the model. We can deduce the following predictions from the vicariance model:

1. If vicariance is the predominant diversification mode, there should be a clustering of distribution areas. We should be able to discern distinct biotic elements.
2. If speciation is mainly due to vicariance, closely related species originate in different areas of endemism and, hence, belong to different biotic elements.
3. If the radiations of some taxa are the result of the same sequence of vicariance events, the cladograms of these taxa should be concordant, which means that species belonging to the same biotic element will have the same topological position in the cladograms.

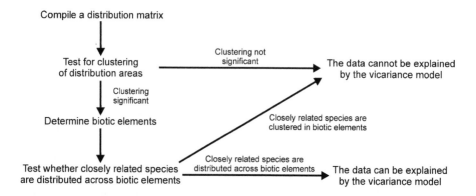

FIGURE 4.1 Flow scheme illustrating the approach used to test whether the data can be explained by the vicariance model.

In the following we describe tests of the first two predictions. The testing approach is illustrated in Figure 4.1.

The first step in any biogeographical analysis is the compilation of a distribution matrix. The taxa and the geographical areas, which shall be investigated, have to be defined. The taxa that shall be examined need not belong to a single clade. However, we recommend combining only taxa with similar dispersal abilities in tests of the vicariance model. If taxa with very different dispersal abilities are combined in the same analysis, existing patterns might be obscured. For example, birds may not be affected by vicariance events that structure the soil fauna. An analysis of a data set including some bird species and some soil organisms might fail to detect clustering of ranges, although the soil fauna was structured by vicariance. On the other hand, the identification of the same biotic elements in analyses of groups of taxa with different dispersal abilities corroborates the generality of the patterns.

The geographical units may be geomorphological regions like islands or drainage basins or arbitrary squares defined by a grid. These units have to be small enough to assure that ranges of species belonging to different biotic elements are not restricted to the same geographical unit. This can be assured by choosing units that are distinctly smaller than the average range size of the examined taxa.

TESTS OF THE VICARIANCE MODEL

Test for Clustering of Distribution Areas

To test the first prediction derived from the vicariance model, namely that there is a clustering of distribution areas, we investigate whether the observed degree of clustering of ranges can be explained by the range size distribution, the varying number of taxa per cell, and the spatial autocorrelation of the occurrences of a taxon alone. Three specifications must be made for the test: a distance measure, a test statistic, and a null model for the generation of sets of ranges (Hausdorf & Hennig, 2003a; Hennig & Hausdorf, 2004). The distribution of the test statistic under the null hypothesis is approximated by a Monte Carlo simulation, which depends on some parameters which are estimated from the data.

Distance Measure

The test for clustering of distribution areas is based on a dissimilarity measure between the ranges of the examined taxa. Originally we used Kulczynski distances:

$$d_K(A,B) = 1 - \frac{1}{2}\left(\frac{|A \cap B|}{|A|} + \frac{|A \cap B|}{|B|}\right) \tag{4.1}$$

$A, B \subseteq R$ denote distribution areas of two taxa that are subsets of the set of geographic units of the study region R. $|A|$ is the number of geographic units of A, $|A \cap B|$ is the number of cells where both taxa are present.

We also developed a generalized version of the Kulczynski distance, the geco coefficient (Hennig & Hausdorf, 2006), that takes the geographic distances between the occurrences of the taxa into account and that is more robust against the pervasive problem of incomplete sampling.

The geco coefficient in general form is defined as follows:

$$d_G(A,B) = \frac{1}{2}\left(\frac{\sum\limits_{a \in A} \min\limits_{b \in B} u(d_R(a,b))}{|A|} + \frac{\sum\limits_{b \in B} \min\limits_{a \in A} u(d_R(a,b))}{|B|}\right) \tag{4.2}$$

d_R can be defined as the geographic distance between geographic units. u is a monotone increasing transformation with $u(0) = 0$. We suggest for geographical distances a transformation u that weights down the differences between large distances. A simple choice of such a transformation is the following:

$$u(d) = u_f(d) = \begin{cases} \dfrac{d}{f \max d_R} & : \ d \leq f \max d_R \\ 1 & : \ d > f \max d_R \end{cases} \quad ,0 \leq f \leq 1 \tag{4.3}$$

That is, u_f is linear for distances smaller than f times the diameter (maximum geographical distance) of the considered region R, while larger geographical distances are treated as "very far away", encoded by $u_f = 1$. $f = 0$ (or f chosen so that $f\max d_R$ is smaller than the minimum nonzero distance in R) yields the Kulczynski distance, and $f = 1$ is equivalent to u chosen as the identity function scaled to a maximum of 1. f should generally be chosen so that $f\max d_R$ can be interpreted as the minimal distance above which differences are no longer meaningful with respect to the judgment of similarity of ranges. We suggest $f = 0.1$ as a default choice, assuming that the total region under study is chosen so that clustering of ranges may occur in much smaller subregions, and that relevant information about a particular unit (e.g., about possible incomplete sampling) can be drawn from a unit which is in a somewhat close neighbourhood compared to the whole range of the region.

Test Statistic

Clustering of ranges means that distances between ranges of the same cluster are small, whereas the distances between ranges of different clusters are large. The variation of distances of a homogeneously distributed set of ranges is expected to be lower, since there is no clear distinction between ranges that belong together and ranges that should be separated.

Based on these considerations, we used the ratio between the sum of the 25% smallest distances (including those within clusters) and the sum of the 25% largest distances (including those between clusters) as test statistic:

$$
T := \frac{\sum_{i \leq 0.25n(n-1)/2} d_{i\,:\,n(n-1)/2}}{\sum_{i \geq 0.75n(n-1)/2} d_{i\,:\,n(n-1)/2}}
\tag{4.4}
$$

If n is the number of ranges, then there are $n(n-1)/2$ distances between ranges. $d_{1:n(n-1)/2} \leq d_{2:n(n-1)/2} \leq \ldots \leq d_{n(n-1)/2:n(n-1)/2}$ denotes the ordered distances. This statistic is expected to be small for clustered data compared to homogeneous data. See also Hennig and Hausdorf (2004) for a comparison of test statistics for clustering including the present proposal.

Null Model

The null model should simulate the case in which all inhomogeneities of the data can be attributed to varying range sizes, to varying numbers of taxa per geographic unit, and to the spatial autocorrelation of the occurrences of a taxon. We developed a null model in which the non-occurrence of range clusters is modelled so that all ranges are generated independently according to the same probabilistic routine. This routine yields ranges such that their cell number distribution approximates the actual distribution of the number of cells per range, the richness distribution of the cells approximates the actual richness distribution of the cells, and the tendency to form disjunct areas is governed by a parameter, which is estimated from the real data set. Single ranges are generated in the following way. First, the number of cells is drawn from the original distribution of range sizes. Then, the first cell is drawn from the set of cells with probabilities proportional to their species richnesses. For all further cells, it is first determined whether a cell should be drawn from the neighbouring cells of the current preliminary range or from its non-neighbours. The disjunction parameter mentioned above is the probability to draw a neighbouring cell. The cell is then drawn from the corresponding subset (neighbours and non-neighbours), again with probabilities proportional to species richnesses. Computational details have been described elsewhere (Hausdorf & Hennig, 2003a; Hennig & Hausdorf, 2004).

DETERMINATION OF BIOTIC ELEMENTS

If there is a significant clustering of distribution areas, biotic elements are determined using model-based Gaussian clustering (MBGC) with noise component as implemented in the software MCLUST (Fraley & Raftery, 1998), because this method provides decisions about the number of meaningful clusters and about ranges that cannot be assigned adequately to any biotic element. MBGC operates on a dataset where the cases are defined by variables of metric scale. Therefore a multidimensional scaling on the matrix of distances between ranges is performed. Furthermore, MCLUST requires an initial estimation of noise (i.e., points that do not fit in any cluster), which is done by the software NNCLEAN (Byers & Raftery, 1998) as suggested by Fraley and Raftery (1998). As tuning constant (number of nearest neighbours taken into account) for NNCLEAN we chose k = number of species/40, rounded up to the next integer. Four MDS dimensions were used.

TEST FOR DISTRIBUTION OF SPECIES GROUPS ACROSS BIOTIC ELEMENTS

After biotic elements are determined, the second prediction of the vicariance model, namely that closely related species belong to different biotic elements, because they originated in different areas of endemism, can be tested. The species are classified according to systematic groups (e.g., subgenera or genera; rows), and biotic elements (columns) in a cross-table. Species belonging to the noise category and species without closely related species represented in the biotic elements are omitted. The vicariance model predicts a uniform distribution of species groups across biotic elements. This corresponds to the cross-table expected under the null hypothesis of independence of rows and columns. We used the chi-square test for independence of rows and columns of the cross-table to test the vicariance model.

The tests as well as the method for the delimitation of biotic elements are implemented in the program package PRABCLUS which is an add-on package for the statistical software R. These programs are available at http://cran.r-project.org.

CASE STUDIES

NORTH-WEST EUROPEAN LAND SNAILS

In a first case study, we investigated distribution data of the north–west European land snails (Hausdorf & Hennig, 2004). Land snails are excellent model organisms for biogeographical studies because of their low dispersal abilities. We digitized distribution maps of 366 north–west European land snail species from Kerney et al., (1983) by using 100-km Universal Transverse Mercator (UTM) grid squares.

For this data set the test statistic T for clustering, the ratio of the sum of the 25% smallest Kulczynski distances to the sum of the 25% largest Kulczynski distances between the ranges, is 0.372. T varied between 0.358 and 0.448 for 1000 artificial data sets generated under the null model (mean 0.412). Thus, T is significantly smaller ($p = 0.004$) for the north-west European land snail species data set than should be expected under the null model. This indicates that the distribution areas of the north-west European land snail species are clustered as predicted by the vicariance model.

We found eight biotic elements and a noise component with PRABCLUS. The partition of the 366 north–west European land snail species is shown in a non-metric multidimensional scaling (Figure 4.2; four dimensions used, stress 26.0%; the solution does not differ from the metrical multidimensional scaling). It can easily be seen that many species cannot be classified into well-separated clusters and were included into the noise component, which contains 21% of the species. This indicates that these species originated by speciation modes other than vicariance or that their ranges were extensively modified by postspeciational range shifts.

One of the eight clusters includes the widespread species occurring almost throughout the study area (33 species). This cluster may not be homogenous, but might include different elements (Holarctic, Palaearctic, European, etc.), which

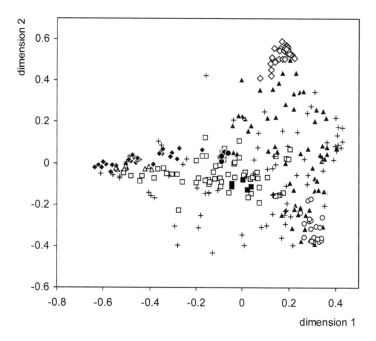

FIGURE 4.2 Non-metrical multidimensional scaling (four dimensions used, only first two dimensions shown; stress 26.0%) of the range data of the north–west European land snail species. Biotic elements found with PRABCLUS: 1/▲ = eastern Alpine–Carpathian element, 2/□ = western Alpine element, ◆ = western European element, 3/O = eastern Alpine element, $y_\bullet^4$ = widespread elements, ◊ = Carpathian element, △ = Pyrenean element, $y_\blacksquare^5$ = south Alpine element, + = noise component.

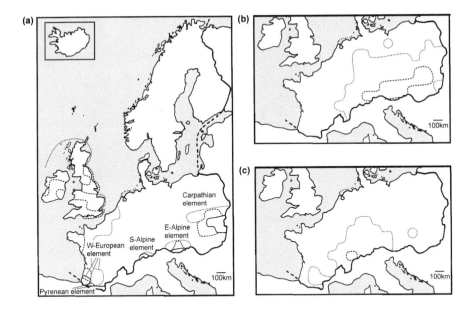

FIGURE 4.3 Distribution maps of regionally restricted biotic elements in north–west Europe. The thick line is the boundary of the study region. Solid line = area where more than 70% of the species of an element are present; dashed line = 50% line; dotted line = 30% line. (a) Distribution of the five restricted regional elements, (b) distribution of the eastern Alpine–Carpathian element, (c) distribution of the western Alpine element.

cannot be distinguished in the present analysis because of the limitation of the study area. The other seven clusters include species with more restricted distributions. Five of these clusters have separate core areas where more than 70% of the species belonging to the biotic element occur. The geographic centres of these biotic elements are in the Pyrenees (16 species), in western Europe (37 species), in the southern Alps (Ticino) (12 species), in the eastern Alps (35 species), and in the Carpathian Mountains (30 species) (Figure 4.3a). The two remaining clusters differ from these restricted biotic elements in the lack of a core area where more than 70% of the species belonging to the biotic element occur. One of the clusters is centred in the eastern Alps–Carpathian Mountains area (68 species; Figure 4.3b), the other in the western Alps (58 species; Figure 4.3c).

Such widespread elements which occupy large regions into which the areas of geographically restricted elements are nested (see also Figure 4.5, where the E-Pyrenean element is nested in the Catalonian element or Figure 4.6–4.7, where the Hermon element is nested in the northern element) might be the result of an expansion of species which originated in one of the areas of endemism or the lack of a response of some species to vicariance events. For example, the areas occupied by the E-Alpine and the Carpathian element are nested in the area occupied by the eastern Alpine–Carpathian element. The eastern Alpine–Carpathian element might include species that originated in the eastern Alps or the Carpathian Mountains and expanded into the other area afterwards, as well as species that were widespread in

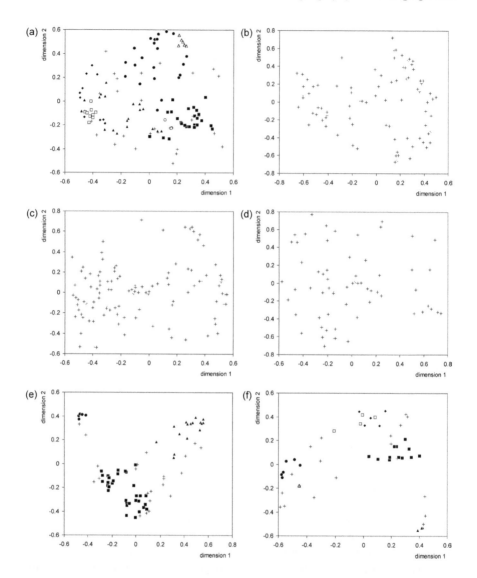

FIGURE 4.4 First two dimensions of the non-metric multidimensional scaling of species range data. The biotic elements are indicated by different symbols for the data sets with a significant clustering (see Table 4.1); + = noise component. (a) Iberian Helicoidea (y■ = Andalusian element;1/▲ = eastern element; 2/● = Basque element; ◆ = Catalonian element; 3/□ = E-Pyrenean element; Δ = Cantabrian element; 4/O = Balearic element); (b) Iberian Helicoidea, species groups with restricted species (no significant clustering); (c) central Aegean land snails (no significant clustering); (d) central Aegean land snails, species groups with restricted species (no significant clustering); (e) Israeli/Palestinian land snails (5/■ = northern element; 6/▲ = southern element; 7/● = Hermon element); (f) Israeli/Palestinian land snails, species groups with restricted species (8/■ = northern element; 9/▲ = Hermon element;10/● = southern element; ◆ = coastal element;11/□ = central element; Δ = southern Judean desert element).

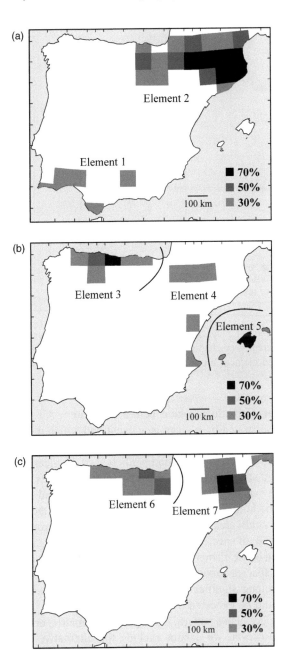

FIGURE 4.5 Distribution maps of biotic elements of the Iberian Helicoidea. The different grey shadings indicate the areas where more than 70, 50, and 30% of the species of an element are present. 1 = Andalusian element; 2 = Catalonian element; 3 = Cantabrian element; 4 = eastern element; 5 = Balearic element; 6 = Basque element; 7 = E-Pyrenean element.

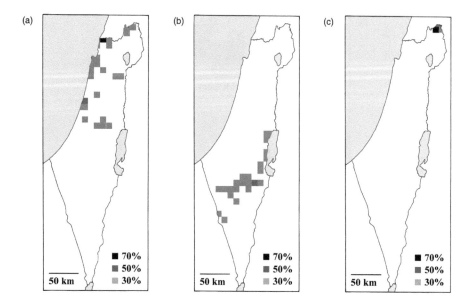

FIGURE 4.6 Distribution maps of biotic elements of the land snail fauna in Israel and Palestine. The different grey shadings indicate the areas where more than 70, 50, and 30% of the species of an element are present. (a) Northern element, (b) Southern element, (c) Hermon element.

the eastern Alpine–Carpathian region before the eastern Alps and the Carpathian Mountains were separated by a vicariance event and which did not respond to that event.

In the next step we tested the second prediction of the vicariance model, namely, that closely related species belong to different biotic elements. For this analysis we omitted the noise category and the cluster of the widespread species, because this cluster may not be homogenous and because it cannot be the result of vicariance events within the study area. We considered subgenera as groups of closely related species. We omitted all species of which there are no closely related species represented in the seven biotic elements with restricted distributions. A total of 161 species belonging to 48 species groups remained. A chi-square test showed that closely related species belong significantly ($p < 0.00001$) more often to the same biotic element than should be expected by chance.

We would expect such a pattern, if peripatric, parapatric, or sympatric speciation were frequent. However, we cannot exclude the alternative explanation that the imprint of vicariance on the distribution patterns has been completely obscured by extensive postspeciational dispersal or regional extinction.

These results may be partly explained by range shifts that were caused by Pleistocene climatic fluctuations. The core areas of the regionally restricted biotic elements (Pyrenees, western Europe, southern Alps, eastern Alps, Carpathian Mountains; Figure 4.3a) correspond well with the centres of sets of nested subsets (Hausdorf & Hennig, 2003b) as well as with supposed glacial refugia of trees (Huntley

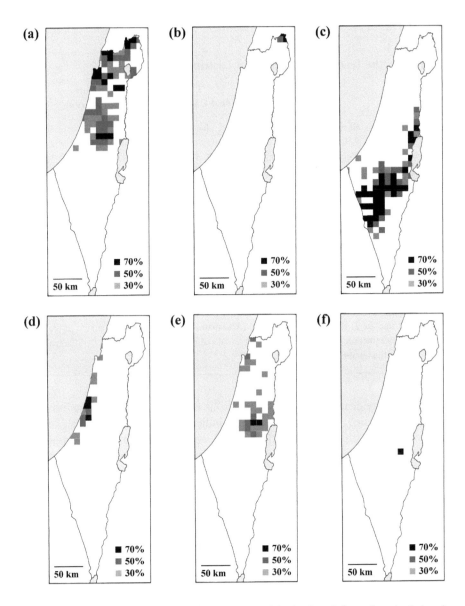

FIGURE 4.7 Distribution maps of biotic elements of the land snail fauna (species belonging to groups with geographically restricted species only) in Israel and Palestine. The different grey shadings indicate the areas where more than 70, 50, and 30% of the species of an element are present. (a) Northern element, (b) Hermon element, (c) southern element, (d) coastal element, (e) central element, (f) southern Judean desert element.

& Birks, 1983). Glacial refugia north of the Mediterranean peninsulas (Iberia, Italy and the Balkans), e.g., at the borders of the Alpine region, have been supposed for many land snails and other invertebrates for a long time (e.g., Poliński, 1928; Zimmermann, 1932; Klemm, 1939, 1974; Holdhaus, 1954; Janetschek, 1956) and

TABLE 4.1
Results of the Tests for Clustering of Distribution Areas

	Number of Species	T (Real Data Set)	T(1000 Monte Carlo Simulations)	P
Iberian Helicoidea, all species	140	0.413	0.401–0.554, mean 0.455	0.009
Iberian Helicoidea, species groups with restricted species	91	0.516	0.436–0.672, mean 0.532	0.353
Aegean land snails, all species	152	0.298	0.270–0.360, mean 0.320	0.058
Aegean land snails, species groups with restricted species	84	0.337	0.308–0.421, mean 0.358	0.072
Israeli/Palestinian land snails, all species	85	0.309	0.304–0.418, mean 0.362	0.004
Israeli/Palestinian land snails, species groups with restricted species	54	0.309	0.266–0.433, mean 0.350	0.046

Note: Test statistic T: ratio of the sum of the 25% smallest geco distances to the sum of the 25% largest distances between the species ranges. p: simulated probability with which T of the real data set is smaller than under the null model.

there is fossil evidence for such refugia (Willis et al., 2000). The core areas of the regionally restricted biotic elements might indicate the positions of such "cryptic northern refugia" (in the sense of Stewart & Lister, 2001) of land snails, especially of forest-dwelling species, which could not survive in the tundra-steppe belt between the North European and the Alpine ice sheets (Hausdorf & Hennig, 2003b, 2004).

MEDITERRANEAN LAND SNAILS

Central Europe is not an ideal region to study vicariance because of the range shifts one can expect due to the Pleistocene climatic fluctuations. Therefore, we examined three distribution data sets of Mediterranean land snails (Hausdorf & Hennig, 2006). The data sets include distribution data of 140 Helicoidea species on the Iberian Peninsula (Puente et al., 1998), the complete land snail faunas of 34 central Aegean Islands (152 species; Hausdorf & Hennig, 2005) and the complete land snail faunas of Israel and Palestine (85 species, data from the database of the Israel National Mollusc Collection in the Hebrew University of Jerusalem; J. Heller, pers. comm., 2002).

In the Iberian Helicoidea and the land snail fauna of Israel and Palestine, the test statistic for clustering of ranges, the ratio of the sum of the 25% smallest geco distances to the sum of the 25% largest geco distances between the ranges, is significantly smaller for the real data set than should be expected under the null model (Table 4.1). Thus, the test indicates that the distribution areas of these Mediterranean land snail species are clustered as predicted by the vicariance model. On the other hand, the test indicates that the distribution areas of the central Aegean land snail species are not significantly clustered. There is no evidence for vicariance in this case.

Diversification by vicariance can only be postulated if there are two or more species of a group with boundaries of their distribution areas in the region under consideration. Distribution patterns originating by vicariance events might be obscured by postspeciational range expansions and non-respondence to vicariance events resulting in widespread species and by extinction resulting in relict species without close relatives. If vicariance is the predominant diversification mode, focussing on species groups with two or more species with boundaries of their distribution areas in the region under consideration should reduce the influence of widespread and relict species on the pattern and result in more clearly delimited biotic elements.

Thus, we performed a second set of analyses in which only species groups with two or more species with boundaries of their distribution areas in the study areas are considered. Contrary to our expectation, the test statistic T, which becomes minimal for highly clustered data, and the probability with which the observed clustering is due to chance increased for the Iberian Helicoidea (91 species) and the central Aegean land snail species (84 species) and remained constant for the land snails from Israel and Palestine (54 species) (Table 4.1). Widespread and relict species do not obscure the clustering of distribution areas expected under the vicariance model. Rather, they contribute to the clustering. Actually, the exclusion of widespread and relict species makes the degree of clustering observed in the Iberian Helicoidea data set also insignificant.

The clustering of the Aegean land snail species and of the Aegean land snail species groups with geographically restricted species is only marginally non-significant. The species which are most strongly clustered in the Aegean data sets are endemics of single islands, especially of the largest islands, Chíos and Sámos. The clustering of these species might be an artefact of the use of islands as geographic units. The tests do not demonstrate that single island endemics are really clustered, because the distribution of these species within the islands is not analysed. Moreover, some widespread, partly even Palaearctic, species are known only from one of the large islands and contribute also to this clustering. Thus, the weak, non-significant clustering in the Aegean data sets should not be considered as evidence for vicariance.

In the next step we determined biotic elements for those data sets for which the previous test indicated that the distribution areas of the examined taxa are significantly clustered. The partition of the species to clusters found with MCLUST is shown in the first two dimensions of non-metric multidimensional scalings in Figure 4.4.

In the MCLUST analysis of the 140 Iberian Helicoidea species (Figure 4.4a; MDS stress 25.2%) 115 species were assigned to seven regionally restricted biotic elements (Figure 4.5), namely, an Andalusian element (30 species), an eastern element (23 species), a Basque element (22 species), a Catalonian element (12 species), an east Pyrenean element (11 species), a Cantabrian element (9 species), and a Balearic element (8 species). 25 species (18%) were included in the noise component.

In the MCLUST analysis of the 85 Israeli/and Palestinian land snail species (Figure 4.4e; MDS stress 17.3%) 62 species were assigned to three regionally restricted biotic elements (Figure 4.6). One biotic element is widespread in northern Israel (33 species); one is widespread in southern Israel (17 species); whereas the third is restricted to the Mount Hermon region (12 species). 23 species (27%) are included in the noise component. If only the 54 Israeli–Palestinian land snail species belonging to groups with geographically restricted species were considered, seven regionally restricted biotic elements were found (MDS stress 9.6%; Figure 4.4f and Figure 4.7). The three elements with most species, the biotic elements centred in northern Israel (12 species), in southern Israel (6 species) and in the Mount Hermon region (6 species), correspond well with the clusters found with the complete data set. Additionally, three regionally restricted biotic elements centred in the coastal region (5 species), in central Israel (4 species), and in the Judean desert (only two species, which are repre-sented in the database from only one grid square, but which are known to be more widespread) were found. The remaining 19 species (35%) are included in the noise component.

A chi-square test showed that closely related Iberian Helicoidea species belong significantly more often to the same biotic element as defined for the complete data set than should be expected by chance ($p < 0.001$). The same is true for the complete Israeli–Palestinian data set ($p = 0.041$).

As already noted for the north–west European land snail fauna, we would expect such a pattern if speciation modes other than vicariance, namely peripatric, parapatric, or sympatric speciation, were predominant. The alternative explanation, that the imprint of vicariance on the distribution patterns has been completely obscured by extensive postspeciational dispersal and/or regional extinction, is less likely for the Iberian Helicoidea and the Israeli–Palestinian land snail fauna than for the north–west European land snail fauna, because these areas were less affected by the Pleistocene glacials than central Europe. Extensive dispersal is less likely between islands of an archipelago like the Aegean islands than between directly neighbouring regions on the mainland like the Iberian Peninsula and Israel and Palestine. We would expect the clustering of ranges to be stronger in the Aegean islands than in the examined continental areas, if vicariance was the predominant diversification mode. Thus, the lack of significant clustering of ranges of land snail species in the central and eastern Aegean Islands is especially strong evidence for the hypothesis that speciation modes other than vicariance, namely, peripatric or parapatric speciation, were predominant. Sympatric speciation is rare at the most, because most closely related species have allopatric ranges.

On the other hand, our tests indicate that the patterns in the Israeli–Palestinian land snail species with restricted distributions meet the predictions of the vicariance model: there is a significant clustering of ranges and the hypothesis that closely related species are homogeneously distributed across biotic element cannot be rejected ($p = 0.669$).

In accordance with several other studies (Endler, 1982a, b; Noonan, 1988; Hausdorf, 1996, 2000; Zink et al., 2000), the case studies show that vicariance is often not the predominant diversification mode and/or that postspeciational range shifts have obscured biogeographic patterns that might have been generated by vicariance.

OTHER BIOGEOGRAPHICAL TESTS OF THE VICARIANCE MODEL

Other biogeographical tests of the vicariance model proposed so far are based on a comparison of area cladograms of different taxa. If the cladogenesis of the taxa occurring in a region is determined by successive vicariance events, the area cladograms of these taxa should be concordant (Rosen, 1976, 1978; Platnick & Nelson, 1978; Nelson & Platnick, 1981; Wiley, 1981, 1988; Humphries & Parenti, 1999; Green et al., 2002). However, the area cladogram approach neglects the problems in identifying and delimiting areas of endemism. Areas of endemism cannot be delimited with biogeographical data alone, if dispersal occurred (Hausdorf, 2002). Moreover, there are hardly any studies in which it has been tested whether there is really non-random congruence between area cladograms of different taxa. In the protocol proposed above, it is tested first whether the distribution areas are clustered at all. If this is the case, biotic elements, which would be a consequence of a fragmentation of an ancestral biota by vicariance events, are determined. Only if at least three biotic elements can be identified and if groups of closely related species are not clustered within elements, but are distributed across elements, is it reasonable to convert taxon cladograms in element cladograms by replacing the names of the taxa by the biotic elements to which they belong (Hausdorf, 2002) and to check whether they are concordant. This test demands extensive phylogenetic analyses of the studied taxa. If the observed degree of matching is not greater than would be expected for independently derived cladograms, there is no evidence for vicariance (Simberloff, 1987).

The proposed tests do not depend on the assumption that there are areas of endemism in any case and that more or less arbitrarily chosen palaeogeographical or geomorphological units correspond to these areas of endemism. Another advantage of this stepwise test protocol is that extensive phylogenetic analyses are not necessary for the first tests. Thus, the distribution data of many taxa can be used to test whether there is a general pattern, whereas in tests based exclusively on comparison of cladograms, only a few taxa for which cladograms are available are considered, and it is not tested whether their distribution areas correspond to a more general pattern.

If the tests show that the predictions of the vicariance model are met, the hypothesis that speciation by vicariance was the predominant speciation mode cannot be rejected. If the predictions of the vicariance model are not met, the situation is more complicated. A negative result might indicate that speciation modes other than vicariance were frequent or that distribution patterns which resulted from vicariance events were obscured by extensive postspeciational dispersal. At present it is difficult to distinguish between these alternatives, because we do not have a null model that can simulate dispersal and extinction in a realistic manner so that we could evaluate the real pattern against patterns, which might result under different scenarios. In any case, the biogeographical data cannot be used to reconstruct the history of the biota and the study area in form of a sequence of vicariance events, if the patterns, which are predicted by the vicariance model, are not found.

Another biogeographical approach to investigate the relative frequencies of different diversification modes is to examine the spatial relationships and the relative size of distribution areas of sister species and sister clades (Wiley, 1981; Wiley & Mayden, 1985; Lynch, 1989; Chesser & Zink, 1994; Barraclough & Vogler, 2000). This approach has been criticized, because the current distribution of a species is not necessarily a reliable indicator of the range at the time of speciation (Chesser & Zink, 1994; Barraclough & Vogler, 2000; Losos & Glor, 2003). Our approach to test the vicariance model is less sensitive to postspeciational range expansions or contractions, because it does not depend on relative range sizes and because biotic elements can still be recognized even if there is some postspeciational dispersal or extinction (Hausdorf, 2002).

Furthermore, a specific vicariance hypothesis can be tested, if there is a hypothesis about the event that might have caused the divergence of two species and if an estimate of their divergence time is available. Then, it can be tested whether the event and the divergence occurred at the same time (e.g., Hedges et al., 1992; Vences et al., 2001; Givnish et al., 2004; Cook & Crisp, 2005; but see Heads, 2005).

CONCLUSIONS

The subjects of biogeography are diverse. Biogeographers examine various patterns in the spatial distribution of organisms or their attributes. There are many different mechanisms that might result in these patterns. If one is interested in clustering of distribution areas and in the vicariance model as possible mechanistic explanation of this pattern, the appropriate units are biotic elements, i.e., groups of taxa with similar ranges. Areas of endemism are inadequate units for biogeographical studies, because they cannot be delimited if dispersal occurred.

It should not simply be presupposed that vicariance was the predominant diversification mode and that there must be areas of endemism in a study region. Rather, it should be tested whether the clustering of ranges is stronger than expected by chance and whether the species of the groups of interest are distributed homogeneously across biotic elements, as predicted by the vicariance model, before one applies methods which presuppose that there are areas of endemism within the study area or that there was no dispersal. These tests will show whether there are distribution patterns expected under the vicariance model in a data set, and they will help to identify those taxa and areas that are most promising for further analyses of vicariance events.

REFERENCES

Barraclough, T.G. & Vogler, A.P. (2000). Detecting the geographical pattern of speciation from species-level phylogenies. *American Naturalist,* 155, 419–434.

Byers, S. & Raftery, A.E. (1998). Nearest neighbor clutter removal for estimating features in spatial point processes. *Journal of the American Statistical Association,* 95, 781–794.

Chesser, R.T. & Zink, R.M. (1994). Modes of speciation in birds: a test of Lynch's method. *Evolution,* 48, 490–497.

Cook, L.G. & Crisp, M.D. (2005). Not so ancient: the extant crown group of Nothofagus represents a post-Gondwanan radiation. *Proceedings of the Royal Society of London, B,* 272, 2535–2544.

Croizat, L., Nelson, G. & Rosen, D.E. (1974). Centers of origin and related concepts. *Systematic Zoology,* 23, 265–287.

Endler, J.A. (1982a). Pleistocene forest refuges: fact or fancy? In *Biological Diversification in the Tropics* (ed. by G.T. Prance), pp. 641–657. Columbia University Press, New York.

Endler, J.A. (1982b). Problems in distinguishing historical from ecological factors in biogeography. *American Zoologist,* 22, 441–452.

Fraley, C. & Raftery, A.E. (1998). How many clusters? Which clustering method? Answers via model based cluster analysis. *Computer Journal,* 41, 578–588.

Givnish, T.J., Millam, K.C., Evans, T.M., Hall, J.C., Pires, J.C., Berry, P.E. & Sytsma, K.J. (2004). Ancient vicariance or recent long-distance dispersal? Inferences about phylogeny and South American–African disjunctions in Rapateaceae and Bromeliaceae based on *ndh*F sequence data. *International Journal of Plant Sciences,* 165, S35–S54.

Green, M.D., van Veller, M.G.P. & Brooks, D.R. (2002). Assessing modes of speciation: range asymmetry and biogeographical congruence. *Cladistics,* 18, 112–124.

Hausdorf, B. (1996). A preliminary phylogenetic and biogeographic analysis of the Dyakiidae (Gastropoda: Stylommatophora) and a biogeographic analysis of other Sundaland taxa. *Cladistics,* 11, 359–376.

Hausdorf, B. (2000). Biogeography of the Limacoidea sensu lato (Gastropoda: Stylommatophora): vicariance events and long-distance dispersal. *Journal of Biogeography,* 27, 379–390.

Hausdorf, B. (2002). Units in biogeography. *Systematic Biology,* 51, 648–651.

Hausdorf, B. & Hennig, C. (2003a). Biotic element analysis in biogeography. *Systematic Biology,* 52, 717–723.

Hausdorf, B. & Hennig, C. (2003b). Nestedness of north-west European land snail ranges as a consequence of differential immigration from Pleistocene glacial refuges. *Oecologia,* 135, 102–109.

Hausdorf, B. & Hennig, C. (2004). Does vicariance shape biotas? Biogeographical tests of the vicariance model in the north-west European land snail fauna. *Journal of Biogeography,* 31, 1751–1757.

Hausdorf, B. & Hennig, C. (2005). The influence of recent geography, palaeogeography and climate on the composition of the fauna of the central Aegean Islands. *Biological Journal of the Linnean Society,* 84, 785–795.

Hausdorf, B. & Hennig, C. (2006). Biogeographical tests of the vicariance model in Mediterranean land snails. *Journal of Biogeography,* 33, 1202–1211.

Heads, M. (2005). Dating nodes on molecular phylogenies: a critique of molecular biogeography. *Cladistics,* 21, 62–78.

Hedges, S.B., Hass, C.A. & Maxson, L.R. (1992). Caribbean biogeography: molecular evidence for dispersal in West Indian terrestrial vertebrates. *Proceedings of the National Academy of Sciences of the USA,* 89, 1909–1913.

Hennig, C. & Hausdorf, B. (2004). Distance-based parametric bootstrap tests for clustering of species ranges. *Computational Statistics and Data Analysis,* 45, 875–895.

Hennig, C. & Hausdorf, B. (2006). A robust distance coefficient between distribution areas incorporating geographic distances. *Systematic Biology,* 55, 170–175.

Holdhaus, K. (1954). Die Spuren der Eiszeit in der Tierwelt Europas. *Abhandlungen der zoologisch-botanischen Gesellschaft in Wien,* 18, 1–493.

Holloway, J.D. & Jardine, N. (1968). Two approaches to zoogeography: a study based on the distributions of butterflies, birds and bats in the Indo-Australian area. *Proceedings of the Linnean Society London,* 179, 153–188.

Humphries, C.J. & Parenti, L.R. (1999). *Cladistic Biogeography*, 2nd ed. Oxford University Press, Oxford.

Huntley, B. & Birks, H.J.B. (1983). *An Atlas of Past and Present Pollen Maps for Europe: 0–13000 Years Ago.* Cambridge University Press, Cambridge.

Janetschek, H. (1956). Das Problem der inneralpinen Eiszeitüberdauerung durch Tiere (ein Beitrag zur Geschichte der Nivalfauna. *Österreichische zoologische Zeitschrift*, 6, 421–506, pl. 1.

Kerney, M.P., Cameron, R.A.D. & Jungbluth, J.H. (1983). *Die Landschnecken Nord- und Mitteleuropas.* Parey, Hamburg and Berlin.

Klemm, W. (1939). Zur rassenmäßigen Gliederung des Genus Pagodulina Clessin. *Archiv fü Naturgeschichte, N.F.,* 8, 198–262, pl. 1.

Klemm, W. (1974). Die Verbreitung der rezenten Land-Gehäuse-Schnecken in Österreich. *Denkschriften der ö terreichischen Akademie der Wissenschaften,* 117, 1–503.

Linder, H.P. (2001). On areas of endemism, with an example from the African Restionaceae. *Systematic Biology,* 50, 892–912.

Losos, J.B. & Glor, R.E. (2003). Phylogenetic comparative methods and the geography of speciation. *Trends in Ecology and Evolution,* 18, 220–227.

Lynch, J.D. (1989). The gauge of speciation: on the frequencies of modes of speciation. In *Speciation and Its Consequences* (ed. by Otte, D. & Endler, J.A.), pp. 527–553. Sinauer, Sunderland.

Mast, A.R. & Nyffeler, R. (2003). Using a null model to recognize significant co-occurrence prior to identifying candidate areas of endemism. *Systematic Biology,* 52, 271–280.

Mayr, E. (1944). Wallace's line in the light of recent zoogeographic studies. *Quarterly Review of Biology,* 19, 1–14.

Morrone, J.J. (1994). On the Identification of Areas of Endemism. *Systematic Biology,* 43, 438–441.

Nelson, G. & Platnick, N. (1981). *Systematics and Biogeography: Cladistics and Vicariance.* Columbia University Press, New York.

Noonan, G.R. (1988). Biogeography of North American and Mexican insects, and a critique of vicariance biogeography. *Systematic Zoology,* 37, 366–384.

Platnick, N.I. & Nelson, G. (1978). A method of analysis for historical biogeography. *Systematic Zoology,* 27, 1–16.

Poliński, W. (1928). Sur certain problèmes du développement morphologique et zoogéographique de la faune des Alpes et des Karpates illustrés par l'étude détailée des Hélicidés du groupe *Perforatella* auct. *Annales Musei zoologici polonici,* 7, 137–229.

Puente, A.I., Altonaga, K., Prieto, C.E. & Rallo, A. (1998). Delimitation of biogeographical areas in the Iberian Peninsula on the basis of Helicoidea species (Pulmonata: Stylommatophora). *Global Ecology and Biogeography Letters,* 7, 97–113.

Rosen, D.E. (1976). A vicariance model of Caribbean biogeography. *Systematic Zoology,* 24, 431–464.

Rosen, D.E. (1978). Vicariant patterns and historical explanations in biogeography. *Systematic Zoology,* 27, 159–188.

Simberloff, D. (1987). Calculating probabilities that cladograms match: a method of biogeographical inference. *Systematic Zoology,* 36, 175–195.

Simpson, G.G. (1977). Too many lines; the limits of the Oriental and Australian zoogeographic regions. *Proceedings of the American Philosophical Society,* 121, 107–120.

Stewart, J.R. & Lister, A.M. (2001). Cryptic northern refugia and the origins of the modern biota. *Trends in Ecology and Evolution,* 16, 608–613.

Szumik, C.A., Cuezzo, F., Goloboff, P.A. & Chalup, A.E. (2002). An optimality criterion to determine areas of endemism. *Systematic Biology*, 51, 806–816.

Szumik, C.A. & Goloboff, P.A. (2004). Areas of endemism: An improved optimality criterion. *Systematic Biology*, 53, 968–977.

Vane-Wright, R.I. (1991). Transcending the Wallace line: do the western edges of the Australian region and the Australian plate coincide? *Australian Systematic Botany*, 4, 183–197.

Vences, M., Freyhof, J., Sonnenberg, R., Kosuch, J. & Veith, M. (2001). Reconciling fossils and molecules: Cenozoic divergence of cichlid fishes and the biogeography of Madagascar. *Journal of Biogeography*, 28, 1091–1099.

Wiley, E.O. (1981). *Phylogenetics. The Theory and Practice of Phylogenetic Systematics*. John Wiley & Sons, New York.

Wiley, E.O. (1988). Vicariance biogeography. *Annual Review of Ecology and Systematics*, 19, 513–542.

Wiley, E.O. & Mayden, R.L. (1985). Species and speciation in phylogenetic systematics, with examples from the North American fish fauna. *Annals of the Missouri Botanical Garden*, 72, 596–635.

Willis, K.J., Rudner, E. & Sümegi, P. (2000). The full-glacial forests of central and southeastern Europe. *Quaternary Research*, 53, 203–213.

Zimmermann, S. (1932). Über die Verbreitung und die Formen des Genus Orcula Held in den Ostalpen. *Archiv fü Naturgeschichte, N.F.*, 1, 1–56, pl. 1–2.

Zink, R.M., Blackwell-Rago, R.C. & Ronquist, F. (2000). The shifting roles of dispersal and vicariance in biogeography. *Proceedings of the Royal Society of London, B* 267, 497–503.

5 Evolution of Specific and Genetic Diversity during Ontogeny of Island Floras: The Importance of Understanding Process for Interpreting Island Biogeographic Patterns

Tod F. Stuessy

ABSTRACT

Islands have long been valued as natural laboratories of plant evolution. Surrounded by water, they invite studies on fascinating and dramatic endemic plant groups, such as the well-known Hawaiian lobelioids and silverswords, or *Aeonium* and *Echium* of the Canary Islands. Because of their physical demarcation and isolation from continental source areas, oceanic islands have been attractive subjects for theoretical and quantitative attempts to understand specific diversity. The now classic island biogeographic theory of the 1960s, which emphasized size of island and distance from source area, allowed predictions of species diversity in context of equilibrium between immigration and extinction. As is well known, however, specific diversity in oceanic islands is determined by many additional factors such as biological characteristics of the immigrants, speciation within and among islands, and ecological heterogeneity in the archipelago. Further, the islands themselves are constantly changing, arising from the sea, followed by erosion, subsidence, and eventual disappearance under water again after 5–6 million years. Specific and genetic diversity in islands, therefore, develop and change in response to these many dynamic physical influences The ontogeny of oceanic island floras can be divided into four phases: (1) arrival and establishment, 0–10,000 years; (2) early development, 10,000 years–3 mya; (3) maturation, 3–5 mya; and (4) senescence and extinction, 5–6 mya. In the first phase, immigrants accumulate and in-coming genetic variation is reduced in comparison to source populations. In the second phase, the number of endemic species increases greatly due to cladogenetic speciation. Genetic variation remains low, as these gene pools are partitioned into rapidly diverging lineages in different ecological zones through adaptive radiation. In anagenetically derived species, however, genetic variation accumulates to levels found in progenitors. In the third phase, erosion and subsidence drastically reduce island surface area and corresponding ecological heterogeneity, causing huge losses in specific (extinction) and genetic diversity. In the fourth and last phase, continued subsidence and erosion lead to total extinction and disappearance of the island. During the last 2,000 years humans also have intervened and impacted island environments through cutting of forests, clearing of land for housing and agriculture, fire, and introduction of invasive plants and animals, all of which have caused additional dramatic loss of both specifi and genetic diversity. Island biogeography can only be addressed, therefore, in context of these many physical, ecological, and historical processes over several millions of years. No equilibrium is ever achieved — it is a slow ontogeny, with initial increase of specific diversity, maintenance and modification of genetic diversity, and subsequent reduction of genetic frequencies, populations, and taxa leading to total extinction.

INTRODUCTION

Oceanic islands are natural laboratories and serve as model systems for studying patterns and processes of plant evolution. High levels of endemism, restriction of gene fl w due to oceanic barriers, and known geological ages combine to make them appealing locations for studying evolution, especially in contrast to more complex continental ecosystems. Darwin (1842) and Wallace (1880) were both attracted to islands, and modern biologists continue to be fascinated by them (Nunn, 1994; Grant, 1998; Stuessy and Ono, 1998; Whittaker, 1998; Emerson, 2002).

Many phylogenetic studies have been done on plants endemic to oceanic islands. Early morphological studies using simple cladistic methodology (Gardner, 1976) have been replaced by those based primarily on DNA sequences (Crawford et al., 1992; Sang et al., 1994; Givnish et al., 1995; Francisco-Ortega et al., 1997; Ganders et al., 2000; Mort et al., 2002). The focus of these investigations has been to determine if the island group is monophyletic, to reveal relationships among island taxa, to assess patterns of adaptive radiation including rates of divergence, and to determine continental progenitors.

In parallel with phylogenetic investigations have gone studies on genetic variation within and among populations of island taxa. The most comprehensive assessments have been made for angiosperms of the Canary Islands (Francisco-Ortega et al., 2000) and the Juan Fernandez Islands (Crawford et al., 2001). Other studies have focused on patterns of genetic variation in island endemics in comparison with mainland progenitors (Wendel & Percival, 1990; Crawford et al., 1993), with special interest on genetic correlates of founding events.

Due to their dramatic nature and geographic simplicity, oceanic islands have also attracted attention for attempts to model species diversity. The now classic theory of island biogeography by MacArthur and Wilson (1967) provided a stimulating view of predicting species diversity based primarily on island size and distance from mainland source areas, and the theory continues to have explanatory power in specifi instances (Schoener et al., 2001). Genetic consequences of the theory have also been explored recently (Johnson et al., 2000; Ricklefs & Bermingham, 2004). We now realize that this theory, in general, is insufficient to explain species diversity (Williamson, 1989; Ricklefs & Bermingham, 2001; Emerson & Oromi, 2005), because it overlooks other important factors such as island ecology, divergent rates of colonization, and intra- and inter-island cladogenetic speciation (especially adaptive radiation). Despite difficulties, however, oceanic islands continue to demand attention, and other, more complex, models of species diversity have been attempted (Stuessy et al., 1998a).

Through most of these valuable studies on islands, there has been an unstated assumption that these well-defined land masses have been reasonably stable over evolutionary time. Much has been written about early stages of island formation and ecological modification (e.g., Carlquist, 1974; Thornton, 1996), but less attention has been given to continued island change. It is obvious, however, that due to subsidence and erosion, plus friable volcanic rocks and soils, oceanic islands are extremely unstable through time, eventually disappearing completely under the ocean in approximately six million years (some last longer, of course, depending upon original island size and subsequent volcanic activity, e.g., Fuerteventura in the Canary Islands, c. 21 mya, García-Talavera, 1998). With geological and environmental ontogeny also comes floristic and vegetational ontogeny. That is, the abiotic changes over time will have a major impact on the ecology of the island and the flora contained within it. The eventual fate of oceanic islands is geological subsidence and extinction of all terrestrial organisms.

During island ontogeny, plant speciation also occurs. The traditional model of speciation in islands has emphasized cladogenesis and adaptive radiation, seen so clearly in well-known groups such as the Hawaiian silverswords (Baldwin & Wessa, 2000; Carlquist et al., 2003) and lobelioids (Givnish et al., 1995, 2004), or *Aeonium* (Crassulaceae; Mes and 't Hart, 1996) and *Echium* (Boraginaceae; Böhle et al., 1996) in the Canary Islands. Recent studies, however, have stressed the importance of simple geographic speciation (or anagenetic speciation; Stuessy et al., 1990), whereby a progenitor arrives in an island and simply diverges through time (see also comments in Gillespie, 2005). This type of speciation has been estimated to account for one fourth of the endemic floras of different oceanic islands/archipelagos, ranging from 7 to 88% depending upon the archipelago (Stuessy et al., 2006). These two different modes of speciation not only lead to very different levels of specifi diversity, but they also yield different genetic consequences (low diversity in the former, much higher in the latter; Pfosser et al., 2005).

In view of the geological, environmental, specific, and genetic changes that take place during ontogeny of islands, therefore, it is necessary when making comparisons between taxa of different island systems, or between taxa of island and continental relatives, to be precise on the island stages that are involved. It makes little sense, for example, to talk about genetic consequences of a founder effect if the populations under consideration are on an island in an advanced phase of its ontogeny. Too many other factors will have influenced levels of genetic variation long after establishment of the founding populations. The purpose of this paper, therefore, is to offer a descriptive hypothesis regarding the arrival and establishment, early development, maturation, and senescence and extinction of specific and genetic diversity in oceanic island archipelagos.

GENERAL ASPECTS OF OCEANIC ISLAND ONTOGENY

GEOLOGICAL ONTOGENY

As with all geological aspects of an active and dynamic planet, oceanic islands also proceed through a geological ontogeny. They are first borne from volcanic activity under the sea, in some cases arising from hypothetical hot spots under tectonic plates (such as Hawaii or Juan Fernandez), or in other cases from volcanic activity along plate junctions. Two cases of newly formed islands in recent time are Krakatau (Ernst, 1908; Thornton, 1996) and Surtsey (Fridriksson, 1975). The formation of an oceanic island takes time, especially if there is secondary volcanic activity upon the original island. Such events can be seen presently on the young island of Hawaii in the Hawaiian archipelago. Eventually, however, volcanic forces stop contributing new land area to the island. Erosion then begins on the landscape, due to wind, water, and wave action. A well-documented case is the Hawaiian chain, because as a U.S. state, it has received extensive geological study (Macdonald et al., 1983). From its origin, the general sequence of island development is the creation of amphitheater-headed valleys, broadening of these valleys, lowering of height of mountains, loss of land area by wave action along the shore of the island, and eventual disappearance under the sea (Carlquist, 1980; Ziegler, 2002). Depending upon the stage of geological ontogeny, therefore, the total size of the island will vary significantl , even dramatically, from its origin. In Masatierra Island in the Juan Fernandez archipelago, we have estimated that up to 95% of the surface area of this island has been lost during the past four million years (Stuessy et al., 2005). In another one million years or so, the island will doubtless be totally submerged.

FLORISTIC ONTOGENY

It is evident that the geological ontogeny sketched above will have a major impact on the island environment, in the form of altering temperature, rainfall, humidity, soil type, etc. The plants upon the landscape, therefore, must adapt to these changes. Some do, and some do not. The first challenge is long-distance dispersal from a source region and establishment into a viable self-sustaining population. This accomplished, the next challenge is proliferation into numerous populations, or by dispersal into a population system within the island, that is then acted upon by natural selection within diverse island habitats. The result of successful adaptation to these circumstances is cladogenesis and adaptive radiation (Schluter, 2000). Groups of closely related species develop in geographic and ecological isolation. These complexes are then impacted by reduction of surface area and habitat, slowly leading to decline of specific diversity (Stuessy et al., 2005). Species are pushed closer together, hybridization may occur, and population numbers and sizes diminish. Extinction is the eventual outcome.

GENETIC ONTOGENY

As new species are formed and are modified as sketched above, changes in the genetic composition of populations also occur. It is convenient to focus on the two major modes of speciation in oceanic islands, cladogenesis and anagenesis, to examine the progress of genetic patterning. It should be remembered that modes of speciation in islands appear more limited than those in continental regions, in the sense that few cases of intra-island polyploidy or dysploidy are known (Carr, 1998; Stuessy and Crawford, 1998). Modes of speciation including these cytological changes, therefore, appear rare in oceanic islands. With both cladogenesis and anagenesis, the immigrants to the island will harbor a greatly reduced portion of the complete range of genetic variation present in progenitor populations. This is the well-known founder effect (Frankham, 1997).

In cladogenesis, this relatively depauperate gene pool becomes fragmented and kept under intense selection in different habitats, resulting in conspicuous morphological differences that are worthy of being called different species (or even genera), but that do not show much genetic divergence among them (Crawford & Stuessy, 1997). The genetic divergence is so minor, in fact, that many island endemic congeners can be easily crossed experimentally (Carr & Kyhos, 1986; Brochmann, 1987).

In anagenesis, however, the process and genetic pattern is substantially different. After the initial founder event that greatly restricts original genetic variation in comparison with progenitor populations, the newly established population builds itself in numbers of individuals. These disperse to other regions of the island. If all populations remain in similar environments, or to put this another way, if the entire island is relatively uniform ecologically (as in Ullung Island, Korea; Pfosser et al., 2005), then little divergence occurs, no cladogenesis, and no adaptive radiation. Results in Ullung Island, which has the highest known level of anagenetically derived endemic species (88%; Stuessy et al., unpubl.), reveal high levels of genetic variation in endemic species of the genus *Acer* (Aceraceae; Pfosser et al., 2002) and *Dystaenia* (Apiaceae; Pfosser et al., 2005). Furthermore, there is no geographic partitioning of this variation; it all behaves as one large population. In contrast to genetic correlates during cladogenesis, therefore, anagenetically derived species gain genetic variation through time and present a very different pattern from those that have resulted from adaptive radiation.

A HYPOTHESIS FOR THE ONTOGENY OF OCEANIC ISLAND FLORAS

With these general perspectives regarding geological, floristic, and genetic ontogenies in oceanic islands, it is possible to advance a general hypothesis for the ontogeny of island floras It is convenient to recognize four phases: (1) arrival and establishment; (2) early development; (3) maturation; and (4) senescence and extinction. This is analogous to ontogenies of individuals as well as of species, as pointed out recently by Levin (2000). This is, after all, the scheme of events befalling nearly all abiotic and biotic structures in the universe. A graphic summary of the hypothesis is presented in Figure 5.1. To these four natural phases must be added a more recent human impact, which also has a major influence (largely negative) on specific and genetic diversity during historical time. For each phase the abiotic and biotic constraining factors and their impact on specific and genetic diversity will be mentioned.

PHASE ONE: ARRIVAL AND ESTABLISHMENT (0–10,000 YEARS)

Constraining Factors

The initial challenge for colonization of a newly formed oceanic island is to successfully disperse from source areas. It has always been understood (Carlquist, 1974, 1981) that much of the flora of oceanic islands must develop initially by long-distance dispersal of propagules. Once in an archipelago, and depending upon subsequent earth events such as lowering of sea level during the Pleistocene or new volcanic activity, vicariant events can also have impact on the origin of new taxa. The original propagules, however, must have had mechanisms to float, raft, loft, or attach outside or inside to birds to reach the new islands. Emphasis on long-distance dispersal has resurged of late (DeQueiroz, 2005), but for oceanic islands there has never been any reasonable alternative.

The abiotic constraints governing the arrival and establishment of propagules have much to do with the size of island and distance from source area, these being the principal factors in the well-known theory of island biogeography (MacArthur & Wilson, 1967). The size of an island represents a target area, increasing landing probabilities as size increases. Likewise, the further away an island is from the source region, the lower the probability of successful colonization. Also important, although not stressed in the theory of island biogeography, is the ecology of the newly formed island. A freshly formed volcanic island is obviously not necessarily an ideal habitat for all immigrants. The more close the habitat of the source region is to that of the new island, the more the chances of successful establishment increase.

The biotic constraints of this first phase deal with the dispersal ability of the propagules and the tenacity of early colonizers. Evidence shows that most plants of oceanic islands arrive by bird transport (Carlquist, 1974), which means that fruits often have barbs, awns, hairs or sticky substances, or less often edible fles y berries. Immigrants that are selfing (i.e., self-pollinating and genetically self-compatible), preferably from only one individual, or have vegetative means of reproduction (e.g., stolons, rhizomes, rooting easily at stem nodes), would be the best candidates for success. A perennial habit would also favor persistence until populations could be built up. The early colonists should possess general-purpose adaptable genotypes and plastic phenotypes to survive the harsh volcanic environment. On the plus side, however, will be the reduction of predators, pests, and competition. Being polyploid would also be an advantage, as this often confers greater range of physiological tolerance (Levin, 2002).

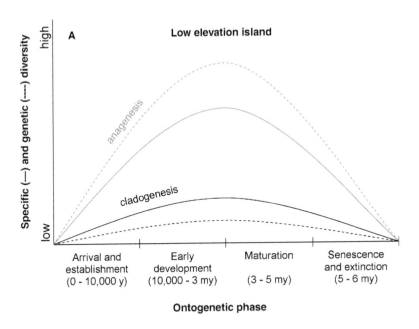

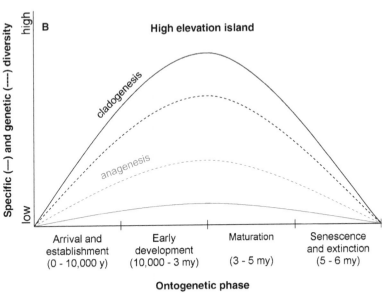

FIGURE 5.1 Models of total amounts of specific (—) and genetic (---) diversity via anagenesis and cladogenesis in the endemic floras during ontogeny in low-elevation (A) and high-elevation (B) oceanic islands.

Impact on Diversity

In this first phase, specific diversity develops from zero. Because only those taxa that are successful dispersers arrive in the new island, the composition of the developing flora is not at all like that of source areas. This new vegetation has been called "disharmonic" (Carlquist, 1974). This is also why some vegetation types in oceanic islands achieve strange appearances (e.g., Mueller-Dombois & Fosberg, 1998) and make them fascinating for biologists and ecotourists. Slowly the ecosystem fills up in this initial phase. That it can happen relatively quickly has been demonstrated by the studies on Krakatau (Ernst, 1908; Thornton, 1996), in which after only 100 years stable forests developed.

Genetic diversity in this first ontogenetic phase will be obviously very restricted in contrast to that of source populations. There is some evidence that additive genetic variation can be generated from only a few immigrants (Naciri-Graven & Goudet, 2003), but it is unclear to what extent this applies to higher plants. In any event, there can be little question that the initial genetic diversity in founding individuals and in subsequent initial populations must be low. The founder effect must be real and substantial. Subsequent introductions can provide added genetic variation, but if the island is very isolated from source areas (e.g., the Hawaiian archipelago), the probabilities of this are low.

The overall evolutionary dynamic in this first phase, therefore, is arrival and establishment into viable populations capable of survival in the new ecosystem. Genetic variation can accumulate to some extent through mutation and recombination, but this will not ordinarily result in speciation in this first ontogenetic phase.

PHASE TWO: EARLY DEVELOPMENT (10,000 YEARS–3 MYA)

Constraining Factors

After arrival and establishment, the next challenge to immigrants is proliferation, in numbers of individuals, numbers and sizes of populations, and eventually in taxic diversity. It is in this second ontogenetic phase that the greatest amount of specifi and genetic diversity accumulates. Once in an island and solidly established, opportunities exist for further diversification Here the ecology of the island matters a great deal, for a varied landscape of sufficient size with heterogeneous habitats provides opportunities for geographic isolation in different habitats and selection, leading to speciation in what is recognized as adaptive radiation. This process leads to the familiar complexes of related species that differ conspicuously in morphological features. The higher the island and therefore the greater the habitat diversity that it harbors, the greater the stimulus for speciation (Stuessy et al., 2006.). Ability to disperse to adjacent islands, if in an archipelago, and continuing adaptive speciation is also important in this phase. There is an interesting dynamic here, in that morphological adaptations for successful dispersal of original colonizers to an island subsequently become a liability on a small island surrounded by ocean. Loss of such structures through time is, therefore, a typical response in many island taxa (Carlquist, 1966). Competition from other established organisms becomes more significant in this phase, as the ecosystem has completely filled in, and each taxon jockeys for a greater share of available resources.

Impact on Diversity

The abiotic and biotic constraints in this second phase lead to the highest levels of specific and genetic diversity during island ontogeny. This is the phase of active speciation via cladogenesis, if the islands's geography and ecology can support it. Large complexes can develop, such as the well-known Darwin's finches (Grant, 1986; Burns et al., 2002; Grant et al., 2004), the Hawaiian silverswords (Asteraceae; Carlquist et al., 2003) and lobelioids (Campanulaceae; Givnish et al., 1995) and *Dendroseris* (Asteraceae; Sanders et al., 1987; Sang et al., 1994) in the Juan Fernandez Islands. The degrees of morphological differentiation can be staggering, as revealed in the Hawaiian archipelago by the genus *Dubautia* (Carr, 1985), in which exists a range from herbs to palmiform trees several meters tall. Despite such morphological adaptations to varied topography and ecology, species are often completely able to cross with each other in garden experiments (e.g., in *Dubautia;* Carr, 1985; Carr & Kyhos, 1986). This demonstrates that the process of adaptive radiation is rapid and that selection is very strong for morphological and physiological adaptations into distinct habitats (Robichaux et al., 1990).

In islands of low elevation and low habitat heterogeneity, however, adaptive radiation cannot operate effectively. In this case, successful population systems proliferate and remain interbreeding, or at least do not diverge, due to more uniform selection pressures over the entire island. Genetic variation accumulates slowly by mutation and recombination, with the entire population system behaving as a single population, or perhaps as slightly differentiated metapopulations. Such a result with AFLP data has been shown (Pfosser et al., 2005) in the endemic *Dystaenia takesimana* in Ullung Island, Korea, where the level of anagenetic speciation is 88% (Stuessy et al., 2006.). This is a low and young island (987 m elevation; 1.8 million years old; Kim 1985) with low habitat heterogeneity. This genetic effect has also been recorded with allozymes in sibling species of *Campanula punctata* in the Izu Islands of Japan (Inoue and Kawahara, 1990).

At the end of this second phase, therefore, the level of genetic variation in anagenetically derived species may reach or surpass that in progenitor populations. With cladogenetically derived species, because each lineage is evolving rapidly and because the initial founding population was itself genetically depauperate, the amount of genetic variation within and between populations and lineages is low (Dodd & Helenurm, 2002), much lower than in comparison to continental relatives, and much lower than in anagenetically derived species. This ontogenetic phase is the most significant for establishing maximal levels of genetic and specific diversity in all populational systems.

PHASE THREE: MATURATION (3–5 MYA)

Constraining Factors

As an island ages, the forces of subsidence and erosion eventually begin to take their toll, resulting in loss of surface area and reduction of habitat. An oceanic island is made up of some resistant rock, often basalts that form the peaks and ridges of an island. The intervening lava and ash, however, are easily weathered and eroded through the constant battering of the island by wave action, rain, wind, temperature fluctuations, land slides, etc. It was always shocking to see the effects of landslides on Masatierra Island in the Juan Fernandez archipelago after the normal spring rains (T. Stuessy, unpubl.), which clear out all plants in their path. Subsidence continues, also, as the tectonic plates upon which the islands sit move slowly over the hotspots and then downward toward their subduction at more distant suture zones. With the Juan Fernandez Islands, for example, this occurs at the western edge of the South American continent. Fire through natural causes also can play a role, although there are no data bearing directly on this point, and it is likely that it is not a major factor. Numerous recent fires set accidentally or deliberately by humans are more threatening to the endemic flora of islands. With regard to biotic constraints in this maturation phase, it is here that landscape resources become ever more limiting, thus heightening competition for survival.

Impact on Diversity

Loss of surface area, reduction of habitat, and increased competition for resources in the maturation phase take their toll on specific and genetic diversity. By this time the major complexes of adaptively radiating endemic species have been formed, and the anagenetic lines have stabilized. It now becomes a mater of holding on and surviving. Continued immigration will continue to occur, obviously, but the closed and tightly constructed ecosystem at this point leaves precious few opportunities for new immigrants to establish themselves. This occurred already largely during the first two phases. Populations are lost and species diversity is lost, perhaps even 25% or more (Stuessy et al., 1998a, 2005), as the island slowly disappears. Geomorphological changes also bring formerly isolated populations into fresh contact, encouraging interspecific hybridization. Correlations of morphological adaptations with original habitat divergence become less clear, as habitats become more generalized; patterns of original adaptive radiation, therefore, become obscured.

With reductions in population size and number in this phase also comes reduction in genetic variation in both cladogenetically and anagenetically derived species. In the former, the reduction may appear even more severe, as the total amount of genetic variation within taxa has always been low. In some groups, the residual genetic variation may begin to approximate that seen in Phase One, i.e., quite low, perhaps even approximating founder-effect levels. Hybridization among populations and taxa brings about more genetic mixing than perceptible in Phase Two. Overall, this phase represents the highest rate of loss of natural specific and genetic diversity.

Phase Four: Senescence and Extinction (5–6 mya)

Constraining Factors

The factors leading to decline of diversity in Phase Three, erosion and subsidence, continue inexorably in this phase until the island disappears under the waves. Island surface area is now seriously reduced, the original peaks and ridges of the island have been worn down, and steep valleys have been broadened out into smooth plains containing a greatly reduced variety of habitats. This, combined with wave action against the now lower sea walls, eventually leads to total encroachment by the sea, leaving only a seamount as remnant.

Impact on Diversity

The drastic reduction of island surface area and habitat in this last phase grinds severely against the remaining specific diversity. Hybridization is now possible among residual surviving endemic congeners as they are crowded even closer on the diminishing landscape. Many endemics are now pushed to extinction. The terrestrial biota eventually becomes totally extinct. Within an archipelago, some taxa may have dispersed successfully to other and younger islands and may still survive. The result within a single island, however, is total organismic loss.

Prior to extinction, loss of number of populations and reduction of size of populations greatly restrict genetic variation, so much so that the end result is similar to that seen from the initial founder effect. Variation within any remaining small populations of cladogenetically derived species becomes very low, as it also does now within populations anagenetically derived. More diverse levels occurring in Phases Two and Three in anagenetic species are now also drained of variation by loss of too many individuals and populations. Extinction obviously eliminates all remaining genetic variation.

HUMAN IMPACT DURING THE PAST 2,000 YEARS

As any biologist well knows, the impact of humans on the biota of oceanic archipelagoes of the world has been dramatic. Humans and their activities bring even further constraining factors on specific and genetic diversity in islands. The list is long and well known, but includes cutting of forests and development (hotels, houses, beaches, roads), fire, and the deliberate introduction of domestic plants and animals. Some of the latter inevitably escape and become feral animals and invasive plants, both of which bring strong pressures against the endemic flora As a general indicator, it has been estimated that only 20% of the native forest remains on Masatierra Island in the Juan Fernandez Islands (Greimler et al., 2002), and this after only 400 years of human intervention (Woodward, 1969; Anderson et al., 2002; Haberle, 2003).

These pressures from human activities slash away also at population number and size, thus further reducing genetic variation within the endemic species. These pressures lead to extinction in many instances, as was documented for the endemic sandalwood in the Juan Fernandez Islands (Stuessy et al., 1998b). The negative impact of the early Hawaiian settlers on the bird fauna of this archipelago is also well documented (Carlquist, 1980; Ziegler, 2002).

Human impacts, because of their scope and intensity, can be significant constraining factors on islands currently in any of the four stages of island floristi ontogeny. One only has to look at the Hawaiian Islands (Ziegler, 2002), belonging to the United States, to see impacts on each island of different geological ages from the oldest (Kauai, 6 mya) to the youngest (Hawaii, < 1 mya). One would probably admit to the greatest human impact on Oahu (4 mya), due mainly to possessing a superior ship harbor (Pearl Harbor) and attractive beaches. Negative pressure on the endemic flora of all these islands, however, has been high, and many endemics of the archipelago are endangered (Wagner et al., 1985).

IMPLICATIONS OF THE HYPOTHESIS

By looking at the ontogeny of specific and genetic diversity within island floras, a number of important points emerge. First, it becomes clear that neither specific nor genetic diversity in islands ever reaches equilibrium, which is the focus of the theory of island biogeography (MacArthur & Wilson, 1967). These levels of diversity simply keep changing during the lifetime of the island (Figure 5.1). Obviously, at a single point in time, there is a level of observable diversity, but this is only fleetin stability. For an island at a particular stage of ontogeny, however, the hypothesis presented here predicts general and relative levels of specific and genetic diversity.

Second, in making comparisons of specific or genetic diversity between islands within or between archipelagos, it is necessary to be aware of the ontogenetic phase of each. It is not fruitful, for example, to discuss genetic components of geographic speciation in a young island still in Phase One with an older island in Phase Three or Four. Teasing out the constraints in each case to make meaningful comparisons is not easy, but awareness is the first step toward greater precision.

Third, in comparisons between continental progenitor and island derivative species pairs (e.g., Wendel & Percival, 1990; Crawford et al., 1993), it is important that the genetic diversity now documented in the island be interpreted in context of the ontogenetic phase in which the island exists (Figure 5.2). Crucial in this regard is correct interpretation of reduced genetic variation in island populations (Frankham, 1997). This is usually attributed to the founder effect, even when the island under concern may be in Phase Three or Four (e.g., Crawford et al., 1993). How to discriminate between low genetic variation due to initial founder effect versus that due to genetic reduction in Phase Three or Four due to loss of number and sizes of populations is a serious challenge. Clearly, more care must be given to these comparisons. Interpreting additional constraints from human activities during historical time is even more challenging.

Native (as opposed to endemic) species have not been discussed directly in this paper, but a few comments may be helpful. Populations resulting from successful immigration and establishment in an island, which, however, never undergo divergence to speciation (either anagenetically or cladogenetically), would remain genetically similar in sharing alleles with progenitor populations. One marked difference, however, would be loss of genetic variation due to the founder effect. Increasing population size in native taxa would slowly result in accumulation of some new genetic variation through mutation and recombination, but it will always be at levels lower than that within anagenetically derived species that have accumulated higher levels of variation (and morphological features; hence their designation as distinct species). Genetic variation within native taxa, however, could approximate that of cladogenetically derived species.

Finally, it is important to stress that the focus of this paper has been on single oceanic islands. Within archipelagos, the possibility for alternative hypotheses increases. The distinction between anagenetic and cladogenetic speciation blurs

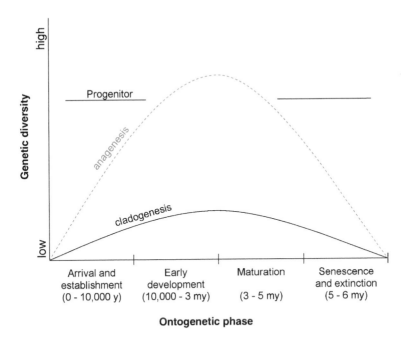

FIGURE 5.2 Model of genetic variation during floristic ontogeny within single endemic species of oceanic islands. Level of genetic variation within hypothetical continental progenitor also shown.

depending upon the nature of the archipelago, in which species will have dispersed simultaneously to several islands or even dispersed serially from one island to another. The ages of the islands can help guide an understanding of which land mass was available for colonization and when. Linear island chains, such as the Juan Fernandez or Hawaiian Islands, allow a general biogeographic stepping-stone model (or "progression", Wagner and Funk, 1995), although back-migrations are known. Mosaic archipelagos, such as the Galapagos Islands, offer many more alternatives for dispersal and patterns of speciation within the island complex, and they are more difficult to interpret in terms of specific and genetic diversity. Despite additional complexities, however, the general ontogenetic phases outlined here would still be significant for helping explain specific and genetic diversity within each of the islands of an archipelago.

ACKNOWLEDGEMENTS

It is a pleasure to thank Malte Ebach for the invitation to participate in the symposium at the Systematics Association meeting in Cardiff, U.K., which resulted in this chapter being prepared. Bernard Hausdorf made several excellent suggestions for improvement of the final manuscript. Appreciation is also expressed to the Austrian National Science Foundation (FWF) for grant No. P14825 for support of these island evolutionary studies.

REFERENCES

Anderson, A., Haberle, S., Rojas, G., Seelenfreund, A., Smith, I. & Worthy, T. (2002). An archaeological exploration of Robinson Crusoe Island, Juan Fernández Archipelago, Chile. In *Fifty Years in the Field. Essays in Honour and Celebration of Richard Shutler Jr.'s Archaeological Career* (ed. by Bedford, S.S.C. & Burley, D.), pp. 239–249. New Zealand Archaeological Association, University of Auckland, Auckland.

Baldwin, B.G. & Wessa, B.L. (2000). Origin and relationships of the tarweed-silversword lineage (Compositae-Madiinae). *American Journal of Botany,* 87, 1890–1908.

Böhle, U.T., Hilger, H.H. & Martin, W.F. 1996. Island colonization and evolution of the insular woody habit in *Echium* L. (Boraginaceae). *Proceedings of the National Academy of Sciences U.S.A.,* 93, 11740–11745.

Brochmann, C. 1987. Evaluation of some methods for hybrid analysis exemplified by hybridization in *Argyranthemum* (Asteraceae). *Nordic Journal of Botany,* 7, 609–630.

Burns, K.J., Hackett, S.J. & Klein, N.K. (2002). Phylogenetic relationships and morphological diversity in Darwin's finches and their relatives. *Evolution,* 56, 1240–1252.

Carlquist, S. (1966). The biota of long-distance dispersal. III. Loss of dispersibility in the Hawaiian flora. *Brittonia,* 18, 310–335.

Carlquist, S. (1974). *Island Biology.* Columbia University Press, New York.

Carlquist, S. (1980). *Hawaii, a Natural History: Geology, Climate, Native Flora and Fauna above the Shoreline,* 2nd ed. Pacific Tropical Botanical Gardens, Honolulu.

Carlquist, S. (1981). Chance dispersal. *American Scientist,* 69, 509–516.

Carlquist, S., Baldwin, B. & Carr, G. (2003). *Tarweeds & Silverswords: Evolution of the Modiinae (Asteraceae).* Missouri Botanical Gardens Press, St. Louis.

Carr, G.D. (1985). Monograph of the Hawaiian Madiinae (Asteraceae): *Argyroxiphium, Dubautia,* and *Wilkesia. Allertonia,* 4, 1–123.

Carr, G.D. (1998). Chromosome evolution and speciation in Hawaiian fl wering plants. In *Evolution and Speciation of Island Plants* (ed. by Stuessy, T.F. & Ono, M.), pp. 5–47. Cambridge University Press, Cambridge.

Carr, G.D. & Kyhos, D.W. (1986). Adaptive radiation in the Hawaiian silversword alliance (Compositae: Madiinae). II. Cytogenetics of artificial and natural hybrids. *Evolution,* 40, 959–976.

Crawford, D.J., Ruiz, E., Stuessy, T.F., Tepe, E., Aqueveque, P., González, F., Jansen, R.J., Anderson, G.J., Bernardello, G., Baeza, C.M., Swenson, U. & Silva O.M. (2001). Allozyme diversity in endemic fl wering plant species of the Juan Fernandez Archipelago, Chile: ecological and historical factors with implications for conservation. *American Journal of Botany,* 88, 2195–2203.

Crawford, D.J. & Stuessy, T.F. (1997). Plant speciation in oceanic islands. In *Evolution and Diversification of Land Plants* (ed. by Iwatsuki, K. & Raven, P.H.), pp. 249–267. Springer Verlag, Tokyo.

Crawford, D.J., Stuessy, T.F., Cosner, M.B., Haines, D.W., Silva O., M. & Baeza, M. (1992). Evolution of the genus *Dendroseris* (Asteraceae: Lactuceae) on the Juan Fernandez Islands: evidence from chloroplast and ribosomal DNA. *Systematic Botany,* 17, 676–682.

Crawford, D.J., Stuessy, T.F., Rodríguez, R. & Rondanelli, M. (1993). Genetic diversity in *Rhaphithamnus venustus,* a species endemic to the Juan Fernandez Islands. *Bulletin of the Torrey Botanical Club,* 120, 23–28.

Darwin, C. (1842). *On the Structure and Distribution of Coral Reefs.* Smith, Elder and Co., London.

DeQueiroz, A. (2005). The resurrection of oceanic dispersal in historical biogeography. *Trends in Ecology and Evolution,* 20, 68–73.

Dodd, S.C. & Helenurm, K. (2002). Genetic diversity in *Delphinium variegatum* (Ranunculaceae): a comparison of two insular endemic subspecies and their widespread mainland relative. *American Journal of Botany,* 89, 613–622.

Emerson, B.C. (2002). Evolution on oceanic islands: molecular phylogenetic approaches to understanding pattern and process. *Molecular Ecology,* 11, 951–966.

Emerson, B.C. & Oromi, P. (2005). Diversification of the forest beetle genus *Tarphius* on the Canary Islands, and the evolutionary origins of island endemics. *Evolution,* 59, 586–598.

Ernst, A. (1908). *The New Flora of the Volcanic Island of Krakatau.* Cambridge University Press, Cambridge.

Francisco-Ortega, J., Santos-Guerra, A., Hinès, A. & Jansen, R.K. (1997). Molecular evidence for a Mediterranean origin of the Macaronesian endemic genus *Argyranthemum* (Asteraceae). *American Journal of Botany,* 84, 1595–1613.

Francisco-Ortega, J., Santos-Guerra, A., Kim, S.-C. & Crawford, D.J. (2000). Plant genetic diversity in the Canary Islands: a conservation perspective. *American Journal of Botany,* 87, 909–919.

Frankham, R. (1997). Do island populations have less genetic variation than mainland populations? *Heredity,* 78, 311–327.

Fridriksson, S. (1975). *Surtsey: Evolution of Life on a Volcanic Island.* Butterworths, London.

Ganders, F.R., Berbee, M. & Pirseyedi, M. (2000). ITS base sequence phylogeny in *Bidens* (Asteraceae): evidence for the continental relatives of Hawaiian and Marquesan *Bidens. Systematic Botany,* 25, 122–133.

García-Talavera, F. (1998). La Macaronesia: consideraciones geológicas, biogeográficas y paleoecológicas. In *Ecologá y Cultura en Canarias* (ed. by Fernández-Palacios, J.M., Bacallado, J.J. & Belmonte, J.A.), pp. 39–63. Organismo Autónomo: Complexo Insular de Museos y Centros, La Laguna.

Gardner, R.C. (1976). Evolution and adaptive radiation in *Lipochaeta* (Compositae) of the Hawaiian Islands. *Systematic Botany,* 1, 383–391.

Gillespie, R.G. (2005). The ecology and evolution of Hawaiian spider communities. *American Scientist,* 93, 122–131.

Givnish, T.J., Montgomery, R.A. and Goldstein, G. (2004). Adaptive radiation of photosynthetic physiology in the Hawaiian lobeliads: light regimes, static light responses, and whole-plant compensation points. *American Journal of Botany,* 9, 228–246.

Givnish, T.J., Sytsma, K.J., Hahn, W.J. & Smith, J.F. (1995). Molecular evolution, adaptive radiation, and geographic speciation in *Cyanea* (Campanulaceae, Lobelioideae). In *Hawaiian Biogeography: Evolution on a Hot Spot Archipelago* (ed. by Wagner, W.L. & Funk, V.A.), pp. 299–337. Smithsonian Institution Press, Washington, DC.

Greimler, J., Lopez S., P., Stuessy, T.F. & Dirnböck, T. (2002). The vegetation of Robinson Crusoe Island (Isla Masatierra), Juan Fernández Archipelago, Chile. *Pacific Science,* 56, 263–284.

Grant, P.R. (1986). *Ecology and Evolution of Darwin's Finches.* Princeton University. Press, Princeton, NJ.

Grant, P.R. (1998). *Evolution on Islands.* Oxford University Press, Oxford.

Grant, P.R., Grant, B.R., Markert, J.A., Keller, L.F. & Petren, K. (2004). Convergent evolution of Darwin's finches caused by introgressive hybridization and selection. *Evolution,* 58, 1588–1599.

Haberle, S.G. (2003). Late Quaternary vegetation dynamics and human impact on Alexander Selkirk Island, Chile. *Journal of Biogeography,* 30, 239–255.

Inoue, K. & Kawahara, T. (1990). Allozyme differentiation and genetic structure in island and mainland Japanese populations of *Campanula punctata* (Campanulaceae). *American Journal of Botany,* 77, 1440–1448.

Johnson, K.P., Adler, F.R. & Cheery, J.L. (2000). Genetic and phylogenetic consequences of island biogeography. *Evolution,* 54, 387–396.

Kim, Y.K. (1985). Petrology of Ulreung volcanic island, Korea — Part 1. Geology. *Journal of the Japanese Association of Mineralogists, Petrologists and Economic Geologists,* 80, 128–135.

Levin, D.A. (2000). *The Origin, Expansion, and Demise of Plant Species.* Oxford University Press, New York.

Levin, D.A. (2002). *The Role of Chromosomal Change in Plant Evolution.* Oxford University Press, Oxford.

MacArthur, R.H. & Wilson, E.O. (1967). *The Theory of Island Biogeography.* Princeton University Press, Princeton.

Macdonald, G.A., Abbott, A.T. & Peterson, F.L. (1983). *Volcanoes in the Sea: the Geology of Hawaii.* University of Hawaii Press, Honolulu.

Mes, T.H.M. & 't Hart, H. (1996). The evolution of growth-forms in the Macaronesian genus *Aeonium* (Crassulaceae) inferred from chloroplast DNA RFLPs and morphology. *Molecular Ecology,* 5, 351–363.

Mort, M.E., Soltis, D.E., Soltis, P.S., Francisco-Ortega, J. & Santos-Guerra, A. (2002). Phylogenetics and evolution of the Macaronesian clade of Crassulaceae inferred from nuclear and chloroplast sequence data. *Systematic Botany,* 27, 271–288.

Mueller-Dombois, D. & Fosberg, F.R. (1998). *Vegetation of the Tropical Pacific Islands.* Springer-Verlag, New York.

Naciri-Graven, Y. & Goudet, J. (2003). The additive genetic variance after bottlenecks is affected by the number of loci involved in epistatic interactions. *Evolution,* 57, 706–716.

Nunn, P.D. (1994). *Oceanic Islands.* Blackwell, Oxford.

Pfosser, M.F., Guzy-Wróbelska, J., Sun, B.-Y., Stuessy, T.F., Sugawara, T. & Fujii, N. (2002). The origin of species of *Acer* (Sapindaceae) endemic to Ullung Island, Korea. *Systematic Botany,* 27, 351–367.

Pfosser, M., Jakubowsky, G., Schlüter, P., Fer, T., Kato, H., Stuessy, T.F. & Sun, B.-Y. (2005). Evolution of *Dystaenia takesimana* (Apiaceae), endemic to Ullung Island, Korea. *Plant Systematics and Evolution,* 256, 159–170.

Ricklefs, R.E. & Bermingham, E. (2001). Nonequilibrium diversity dynamics of the Lesser Antillean avifauna. *Science,* 294, 1522–1524.

Ricklefs, R.E. & Bermingham, E. (2004). Application of Johnson et al.'s speciation threshold model to apparent colonization times of island biotas. *Evolution,* 58, 1664–1673.

Robichaux, R.H., Carr, G.D., Liebman, M. & Percy, R.W. (1990). Adaptive radiation of the silversword alliance (Compositae: Madiinae): ecological, morphological, and physiological diversity. *Annals Missouri Botanical Garden,* 77, 64–72.

Sanders, R.W., Stuessy, T.F., Marticorena, C. & Silva O.,M. (1987). Phytogeography and evolution of *Dendroseris* and *Robinsonia,* tree Compositae of the Juan Fernandez Islands. *Opera Botanica,* 92, 195–215.

Sang, T., Crawford, D.J., Kim, S.-C. & Stuessy, T.F. (1994). Radiation of the endemic genus *Dendroseris* (Asteraceae) on the Juan Fernandez Islands: evidence from sequences of the ITS regions of nuclear ribosomal DNA. *American Journal of Botany,* 81, 1494–1501.

Schluter, D. (2000). *The Ecology of Adaptive Radiation.* Oxford University Press, Oxford.

Schoener, T.W., Spiller, D.A. & Losos, J.B. (2001). Natural restoration of the species-area relation for a lizard after a hurricane. *Science* 294, 1525–1528.

Stuessy, T.F. & Crawford, D.J. (1998). Chromosomal stasis during speciation in angiosperms of oceanic islands. In *Evolution and Speciation of Island Plants* (ed. by Stuessy, T.F. & Ono, M.), pp. 307–324. Cambridge University Press, Cambridge.

Stuessy, T.F., Crawford, D.J. & Marticorena, C. (1990). Patterns of phylogeny in the endemic vascular flora of the Juan Fernandez Islands, Chile. *Systematic Botany,* 15, 338–346.

Stuessy, T.F., Crawford, D.J., Marticorena, C. & Rodríguez, R. (1998a). Island biogeography of angiosperms of the Juan Fernandez archipelago. In *Evolution and Speciation of Island Plants* (ed. by Stuessy, T.F. & Ono, M.), pp. 307–324. Cambridge University Press, Cambridge.

Stuessy, T.F., Greimler, J. & Dirnböck, T. (2005). Landscape modification and impact on specific and genetic diversity in oceanic islands. *Biologiske Skrifter,* 53, 89–101.

Stuessy, T.F., Jakubowsky, G., Salguero Gómez, R., Pfosser, M., Schüter, P.M., Fer, T., Sun, B.-Y. & Kato, H. (2006). Anagenetic evolution in island plants. *Journal of Biogeography,* 33, 1259–1265.

Stuessy, T.F. & Ono, M. (1998). *Evolution and Speciation of Island Plants.* Cambridge University Press, Cambridge.

Stuessy, T.F., Swenson, U., Marticorena, C., Matthei, O. & Crawford, D.J. (1998b). Loss of plant diversity and extinction on Robinson Crusoe Islands, Chile. In *Rare, Threatened and Endangered Floras of Asia and the Pacific Rim* (ed. by Peng, C.-I. & Lowrey, P.P. II), pp. 243–257. Institute of Botany, Academia Sinica Monograph Series No. 16, Taipei.

Thornton, I. (1996). *Krakatau. The Destruction and Reassembly of an Island Ecosystem.* Harvard University Press, Cambridge, MA.

Wagner, W.L. & Funk, V.A. (1995). *Hawaiian Biogeography: Evolution on a Hot Spot Archipeliago.* Smithsonian Institution Press, Washington, D.C.

Wagner, W.L., Herbst, D.R. & Yee, R.S.N. (1985). Status of the native fl wering plants of the Hawaiian Islands. In *Hawai'i's Terrestrial Ecosystems: Preservation and Management* (ed. by Stone, C.P. & Scott, J.M.), pp. 23–74. University of Hawaii, Honolulu.

Wallace, A.R. (1880). *Island Life, or the Phenomena and Causes of Insular Faunas and Floras Including a Revision and Attempted Solution of the Problem of Geological Climate.* Macmillan and Co., London.

Wendel, J.F. & Percival, A.E. (1990). Molecular divergence in the Galapagos Islands-Baja California species pair, *Gossypium klotzschianum* and *Gossypium davidsonii* (Malvaceae). *Plant Systematics and Evolution,* 171, 99–116.

Whittaker, R.J. (1998). *Island Biogeography: Ecology, Evolution, and Conservation.* Oxford University Press, Oxford.

Williamson, M. (1989). The MacArthur and Wilson theory today: true but trivial. *Journal of Biogeography,* 16, 3–4.

Woodward, R.C. (1969). *Robinson Crusoe's Island: A History of the Juan Fernandez Islands.* University of North Carolina Press, Chapel Hill.

Ziegler, A.C. (2002). *Hawaiian Natural History, Ecology, and Evolution.* University of Hawaii Press, Honolulu.

6 Event-Based Biogeography: Integrating Patterns, Processes, and Time

Isabel Sanmartín

ABSTRACT

Cladistic biogeography searches for general patterns of distribution among multiple organisms as evidence of a common biogeographic history, but does not explicitly consider the evolutionary processes that created such patterns or the timing of divergence of the organisms studied. This makes biogeographic results often difficul to interpret. Here, I review recent methodological developments on biogeographic analysis based on the recognition of processes ("the event based approach"), which have led to an extraordinary revolution in biogeographic studies. Event-based methods explicitly include all biogeographic processes into the analysis as possible explanations for the observed biogeographic pattern. Each process (vicariance, duplication, extinction, and dispersal) is associated with a cost inversely related to its likelihood. The optimal biogeographic reconstruction is the one that minimizes the total cost of the implied events (most parsimonious). Significance of results can be assessed by comparing them with those derived from random data sets under the null hypothesis that distributions are not phylogenetically constrained. I compare two event-based methods, dispersal-vicariance analysis and parsimony-based tree fit ting in relation to alternative biogeographic scenarios: hierarchical vs. reticulate, widespread vs. endemic, etc. I describe recent studies in which event-based methods, combined with large data sets of phylogenies and estimates of divergence times, have been used to test large-scale biogeographic patterns on the Holarctic and Southern Hemisphere biotas. Despite claims by cladistic biogeography that dispersal is a random, stochastic event, these studies show that dispersal can also generate congruent distribution patterns among multiple taxa if coordinated in direction, e.g., by prevailing winds and ocean currents ("concerted dispersal"). They also suggest major biogeographic differences between animals and plants, with plant patterns more influenced by recent dispersal.

INTRODUCTION

Historical biogeography aims to infer the distribution history of biotas and to identify the causal factors or processes that have shaped those distributions over time. This discipline plays a crucial role in our efforts to understand how present-day biodiversity has been generated. It addresses such fascinating questions as: Why are organisms distributed where they are today? Where did the fauna of a continent originate? Which processes are responsible for the spatial distribution of plants and animals through time? Biogeographic reconstructions can be used, for instance, to identify *relict* biotas whose long-term survival is threatened, or to predict how flora and faunas will respond to climatic change.

So far, however, the contribution of historical biogeography to conservation decisions has been very little, hampered by methodological controversies and the inadequacy of existing methods to deal with the complexity of real data. Traditionally, biogeographers have explained disjunct (allopatric) distributions by two alternative historical processes: dispersal of the species ancestor across a pre-existent barrier (e.g., a mountain chain, or an ocean), or vicariance caused by the fragmentation of a previously contiguous landmass by a new barrier (e.g., the uplift of a mountain chain, or opening of an ocean). Vicariance is usually assumed to be the primary explanation. Since virtually any distribution pattern can be explained by dispersal, dispersal hypotheses are presumably resilient to falsification (Morrone & Crisci, 1995). On the other hand, vicariance hypotheses can be tested by searching for congruence between the phylogenetic and distribution patterns of different organisms: taxa exhibiting the same phylogenetic and distribution pattern are assumed to have shared a common biogeographic history (i.e., they were part of the same biota and became affected by the same vicariant events). Dispersal, by contrast, was thought to be a rare and random phenomenon and could never be a general explanation for congruence among patterns (Croizat et al., 1974; Craw, 1982; Heads, 1999; Humphries, 2001).

Under the premise that "Earth and life evolve together", cladistic biogeography (Platnick & Nelson, 1978; Nelson & Platnick, 1981; Brooks, 1985; Wiley, 1988; Humphries & Parenti, 1986) was born in the 1980s, as a result of the fusion of cladistic systematics (Hennig, 1966) with the theory of plate tectonics and the concept of vicariance (Croizat, 1964; Brundin, 1966). It represented a huge leap forward over the narrative dispersal scenarios that had dominated biogeography until then (Darlington, 1957) because, for the first time, it provided a method to analyze historical biogeographic patterns quantitatively.

The aim of cladistic biogeography is to identify general patterns in the relationships among areas of endemism among multiple organisms, as evidence of a common biogeographic history. The analysis starts with the construction of *taxon-area cladograms* (TACs) in which the terminal taxa in a phylogeny are replaced by the areas in which they occur today. Comparing TACs of different organisms occurring in the same region may help us to infer a common biogeographic pattern, which can be represented in a *general area cladogram* (GAC), i.e., a branching diagram that expresses the relationships among the areas based on their shared biotas. This general pattern is interpreted as the result of "tectonic history" or vicariance, a sequence of common geological events that affected the evolution of many organisms simulta-neously (e.g., uplift of a mountain chain). The most common cladistic biogeographic methods are Brooks parsimony analysis (BPA, Brooks, 1985, 1990; Wiley, 1988; Brooks et al., 2001), component analysis (Nelson & Platnick, 1981; Humphries & Parenti, 1986; Page, 1990; Ebach et al., 2003), subtree analysis (Nelson & Ladiges, 1996, 1997, 2001) and tree reconciliation (Page, 1994).

Ronquist (1995, 1997, 1998a, b, 2003a) referred to these cladistic methods as "pattern-based" because they focus on finding general patterns of area relationships (GACs), without distinguishing among the evolutionary processes that created such patterns. Although vicariance is assumed in these models, other biogeographic processes such as dispersal or extinction are not incorporated directly into the analysis but considered *a posteriori* (Wiley, 1988), or through *ad hoc* assumptions (Page, 1994), in interpreting any incongruence between the general area cladogram and the individual patterns. However, because any case of incongruence can be explained by several different combinations of events, the choice of a specific set of events that could explain the observations is left to the investigator (Wiley, 1988). Just which biological process is responsible for the incongruence is not clear from the reconstruction (see Page, 1990, and Ronquist & Nylin, 1990, for critiques).

To focus exclusively on vicariance might also be unfounded. For example, many molecular studies support dispersal as the main explanation for the observed distri-butions, because the organism studied is simply too young to have been affected by plate tectonics (Baum et al., 1998; Waters et al., 2000; Buckley et al., 2002). Moreover, despite claims by cladistic biogeography that dispersal is a random, stochastic event unable to produce general patterns (Heads, 1999; Humphries, 2001), recent biogeographic studies based on multiple organism phylogenies (Winkworth et al., 2002; Sanmartín & Ronquist, 2004) have shown that dispersal can indeed produce congruent distribution patterns in the same way as vicariance when it takes place repeatedly in the same direction between the same areas, driven by prevailing winds and ocean currents ("concerted dispersal"). For example, the sister-group relationship between Australia and New Zealand found in many plant groups has been attributed to long-distance dispersal across the Tasman Sea, mediated by the West Wind Drift (Muñoz et al., 2004; Sanmartín & Ronquist, 2004). Moreover, paleogeographic reconstructions indicate that some regions present a complex geo-logical history in which areas have split and fused again in different configuration through time (e.g., the Holarctic region). This has allowed biotas to repeatedly merge and fragment as dispersal barriers disappeared and reappeared in time, resulting in a "reticulate" biogeographic history which cannot be represented into a hierarchical area cladogram (Enghoff, 1995; Sanmartín et al., 2001; Sanmartín, 2003).

Another problem of cladistic methods, discussed by Donoghue & Moore (2003), is that they do not directly incorporate information on divergence times in the reconstruction. This makes it difficult to distinguish between "shared biogeographic history" (i.e., topological and temporal congruence among patterns) and "pseudocon-gruence", when two or more groups show the same pattern but diversified at different times (Cunningham & Collins, 1994). Finally, once a general biogeographic pattern has been deduced, how can we test whether it explains the distributions significantl better than expected by chance, and how can we evaluate its correctness/likelihood

in relation to alternative scenarios? Contrary to what is claimed by cladistic biogeography (Platnick & Nelson, 1978), incongruent area cladograms are not appropriate
tests to falsify biogeographic hypotheses (Simberloff et al., 1981; Andersson, 1996).
We clearly need statistical tests to assess the significance of biogeographic hypotheses in historical biogeography (Seberg, 1988; Altaba, 1998).

In recent years, new methods of biogeographic analysis have been developed that
explicitly incorporate all biogeographic processes into the analysis as possible explanations of the observed pattern, rather than just focusing on vicariance, the so-called
"event-based approach" (Page, 1995; Page & Charleston, 1998; Ronquist & Nylin,
1990; Ronquist, 1994, 1995, 1997, 1998a, b, 2003a; Sanmartín & Ronquist, 2002).
Event-based methods specify the ancestral distributions and the biogeographic processes responsible directly in the reconstruction, so no *a posteriori* interpretation is
necessary. Each type of process (e.g., vicariance, sympatric speciation, extinction, and
dispersal) is assigned a cost that should be inversely related to the likelihood of that
event occurring in the past: the more likely the event, the lower the cost. The optimal
biogeographic reconstruction is found by searching for the reconstruction that minimizes the total cost of the implied events (Ronquist, 2003a). This minimum-cost
reconstruction is also the most likely (most parsimonious) explanation for the origin
of the pattern being analyzed. These methods are thus often called "event-based
parsimony methods". The best-known event-based methods are maximum cospeciation
(Page, 1995; Page & Charleston, 1998; Page, 2003), implemented in the computer
program TreeMap (Page, 1995); dispersal-vicariance analysis (Ronquist, 1997; Sanmartín et al., 2001), implemented in the program DIVA v. 1.0 (Ronquist, 1996); and,
recently, parsimony-based tree fitting (Ronquist, 2003a; Sanmartín & Ronquist, 2002,
2004), implemented in the program TreeFitter 1.2 (Ronquist, 2003b).

A common critique against event-based methods is that if the (cost) model is
wrong, the biogeographic inference would be wrong. The same argument has been
argued against model-based methods, such as maximum likelihood or Bayesian
inference, in phylogenetic analysis. However, "ignoring process in the formulation
of a method does not make it more objective; the method's performance is still
determined by the nature of the evolutionary processes being studied" (Ronquist,
2003a). By making explicit the connection between biogeographic processes and
the distribution patterns they generate, event-based methods can be used to compare
alternative process models/biogeographic scenarios (e.g., exclusively vicariance,
dispersal-vicariance, only dispersal, etc.) Moreover, unlike cladistic biogeography,
event-based methods allow us to analyze each group's individual history separately
before searching for general biogeographic patterns. This gives us a chance to detect
multiple, incongruent histories (Sanmartín et al., 2001) and to determine whether
common patterns can be ascribed to the same processes (McDowall, 2004). For
example, each node in the individual phylogenies could be associated with a time
estimate, based on molecular or fossil evidence, to help us distinguish between
pseudocongruence and shared biogeographic history. Finally, event-based methods
allow us to test the significance of the inferred patterns by comparing the observed
data with those expected by chance under the null hypothesis that organism distributions are not phylogenetically conserved (Ronquist, 2003a; Sanmartín & Ronquist,
2004; see also Page, 1994).

Although event-based methods have gained in popularity in recent years (Voelker, 1999; Sanmartín, 2003; Donoghue & Smith, 2004, Vilhelmsen, 2004), there is still confusion in the literature about their methodology and implied assumptions (Voelker, 1999; Brooks & McLennan, 2001; Crisp & Cook, 2004). In this chapter, I review the event-based approach (for a more complete discussion, see Ronquist, 2003a) and compare two event-based methods, dispersal-vicariance analysis and parsimony-based tree fitting, in relation to alternative biogeographic scenarios. I also illustrate these methods with several case studies in which they have been used to study complex biogeographic patterns (both dispersal and vicariance) on the Holarctic and Southern Hemisphere biotas (Sanmartín et al., 2001; Sanmartín, 2003; Donoghue & Smith, 2004; Sanmartín & Ronquist, 2004).

PARSIMONY-BASED TREE FITTING

Event-based biogeographic methods rely on explicit models with states (distributions of terminals) and transition events between states (biogeographic processes) (Sanmartín & Ronquist, 2002). The model involves: (a) the phylogeny of the species, (b) the associated geographic distributions, (c) a general area cladogram (GAC) describing the relationships among the areas occupied by the species, and (d) a biogeographic model specifying the types of events that can produce the distributions and the cost assignments to each event in the model (Ronquist, 2003a).

The most common model is the four-event-based model (Page, 1995; Ronquist, 2003a) with four biogeographic processes:

- *vicariance* (v): allopatric speciation in response to a general dispersal barrier affecting many organisms simultaneously
- *duplication* (d) or speciation within the area: sympatric speciation or alternatively, allopatric speciation in response to a temporary dispersal barrier affecting a single organism lineage
- *extinction* (e): disappearance of a lineage from an area where it is predicted to occur
- *dispersal* (i): colonization of a new area by crossing a pre-existent barrier followed by speciation.

In this model, dispersal and extinction events are always associated with speciation. This means that they must leave at least one observable descendant at the original host area; otherwise, these events cannot be traced in the phylogeny, such as, for example, in complete dispersals and extinctions (Page, 1995; Ronquist, 2003a).

Once each event in the model is assigned a cost, it is possible to fit an organism phylogeny and its associated geographic distributions to any given area cladogram. This is the *reconstruction* problem (Ronquist, 2003a). The procedure can be compared to the optimization of a character transformation matrix onto a phylogeny (or a parasite tree onto a host tree) using phylogenetic parsimony analysis (Ronquist & Nylin, 1990; Page, 1995). Figure 6.1 shows a simple example to illustrate this. Consider a TAC with fi e taxa distributed in four areas (Figure 6.1a) and a GAC describing the relationships among those areas (Figure 6.1b). The TAC is fitted to

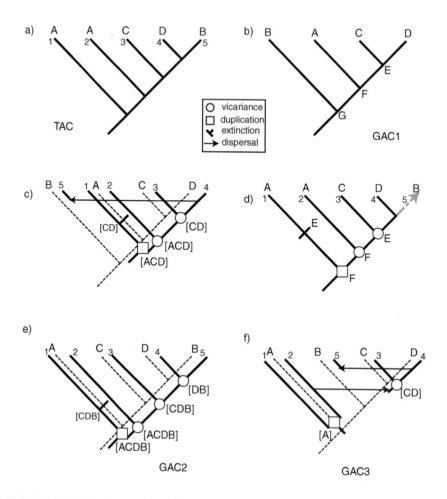

FIGURE 6.1 A simple example to illustrate parsimony-based tree fitting in historical bioge-ography. (a) Taxon-area cladogram (TAC) with fi e taxa distributed in four areas. (b) General area cladogram (GAC1) describing the relationship between the areas. (c) Trackogram illus-trating the most parsimonious reconstruction of the distribution history of the organisms by showing the TAC on top of the GAC (dashed line). (d) The same reconstruction illustrated as an "event-mapping", in which the biogeographic events are specified on the TAC. (e–f) Two other GACs showing alternative resolutions of the relationships between the areas occupied by the TAC, and their corresponding event-based reconstructions (see text).

the GAC to produce a reconstruction of the biogeographic history of the organisms according to the GAC (Figure 6.1c–d). The reconstruction can be illustrated either as a trackogram displaying the organism phylogeny on top of the area cladogram (Figure 6.1c), or as an event-mapping, in which the biogeographic events are spec-ified on the TAC (Figure 6.1d), with symbols denoting the four kinds of events (Page, 1995; Sanmartín & Ronquist, 2002). The cost of the reconstruction is found by simply summing over the cost of the biogeographic events that are required to explain the observed distributions according to the area cladogram. For example,

GAC1 (Figure 6.1c–d) requires a basal duplication event in ancestral area F [ACD] giving rise to two widespread lineages, one of which goes extinct in area E [CD], whereas the second lineage undergoes successive vicariance coupled with extinction events followed by secondary dispersal from D to B. The cost of this reconstruction is therefore: c = 1d + 2v + 1e + 1i).

TESTING SIGNIFICANCE

Any GAC, however, can be fitted to a particular TAC, providing that enough events are postulated. Once a general biogeographic pattern (GAC) has been found, we want to test whether it explains the distributions significantly better than expected by chance. How well supported is our reconstruction? The most common method to evaluate statistical significance in biogeography is to compare the observed cost (i.e., the cost of fitting the GAC and the original TAC) with a distribution of cost values expected by chance under the null hypothesis that there is no "historical association" between the taxon cladogram and the area cladogram, that is, the distributions in the TAC are random and not determined by the phylogeny ("phylogenetically constrained", Ronquist, 2003a). Assume that we generate 100 random data sets in which the terminal distributions in the TAC, the terminal areas in the GAC, or both, have been randomly permuted and calculate the cost of the reconstructions. The significance of the fit is determined by comparing the observed cost with the distribution of random cost values (the "expected cost"). If less than 5% of the random data sets had a cost less than or equal to that of the observed data, we can reject the null hypothesis of "no phylogenetic constraint" at the 0.05 significance level. The p value is therefore the proportion of random data sets that fit the area cladogram better or as well as the original data set. Alternatively, one can permute the topology of the area cladogram or the organism phylogeny instead of the terminals (Page, 1995), but this can be problematic because it requires the choice of an appropriate tree distribution and depends on the shape of the original trees (e.g., balanced/unbalanced) (Ronquist, 2003a).

FINDING THE OPTIMAL COST ASSIGNMENTS

From the discussion above, it can be deduced that the most important problem in parsimony-based tree fitting is to find the cost for each type of biogeographic event. In maximum cospeciation (Page, 1995), also termed "maximum vicariance" (Ronquist, 2003a), this is solved by searching for the reconstruction that maximizes the number of inferred vicariance events. This is equivalent to giving a negative cost (a benefit to vicariance and zero cost to all the other events. A problem of this approach is that it can give reconstructions in which vicariance events that are weakly supported by the data are mixed with well-supported vicariance events, or reconstructions with an unrealistic large number of extinctions and dispersals (Ronquist, 2003a). Instead, Ronquist (2003a) proposed an alternative procedure to set the cost assignments based on maximizing the likelihood of finding phylogenetically conserved distribution patterns. Assume that we use permutation tests of distributions to test for significanc of results as described above. The optimal cost assignments are those that minimize the probability of the permuted data sets having a total cost lower than that of the observed data (Sanmartín & Ronquist, 2004). Examination of simulated data sets (Ronquist, 2003a) showed that a combination of arbitrarily low cost for vicariance and duplication events and a higher cost for extinction and dispersal events gives us the best chances to find historically conserved distributions (low p value) under a wide range of conditions. This occurs because both vicariance and duplication events are phylogenetically constrained processes, whereas dispersal and extinction are not. In duplication and vicariance events, the descendants remain in the ancestral host areas: either both descendants "inherit" the whole ancestral range (duplication or sympatric speciation), or each receives a subset so that the sum of the descendant distributions is equal to the range of the ancestor (vicariance). In other words, duplication and vicariance events generate distribution patterns that are constrained by the phylogeny, i.e., limited by ancestor/descendant relationships. By contrast, dispersal and extinction events are largely independent of phylogeny: the distributions are the result of an independent event, i.e., the colonization of a new area by one of the descendants (dispersal) or the disappearance of the lineage from part of the ancestral area (extinction). Therefore, these events do not produce phylogenetically constrained distribution patterns, but can wipe out their traces if they are not penalized in the reconstruction by giving them a higher cost. In addition, Sanmartín & Ronquist (2002) showed that a low extinction cost with respect to dispersal minimizes the importance of absence data (missing areas) in event-based biogeographic reconstruction. Therefore, extinction carries a lower cost than dispersal in parsimony-based tree fitting. Also, extinction can be considered equivalent to losses or reversals in character evolution, which are probably more frequent in an evolutionary sense than multiple convergences or parallelisms (Cook & Crisp, 2005), the equivalent to dispersal events in biogeography. In the following examples, I will use the default cost assignments of TreeFitter v. 1.2: 0.01 for vicariance and duplication, 1.0 for extinction, and 2.0 for dispersal, which are within the optimal range proposed by Ronquist (2003a).

SEARCHING FOR THE BEST AREA CLADOGRAM

Once the event costs are fi ed, we can search for the best area cladogram for the observed TAC. This is the *estimation* problem (Ronquist, 2003a). The original TAC is fitted in turn to each possible GAC (every combination of the areas). The optimal GAC is the one that gives the most parsimonious reconstruction of the observed distributions in the TAC (i.e., the reconstruction of minimum cost). For example, Figure 6.1 shows three different GACs specifying alternative relationships among the areas. According to our cost assignments, the cost of GAC1 (Figure 6.1c) is c = (0.01 _ d) + (0.01 _ 2v) + (1 _ e) + (2 _ i) = 3.03, whereas the cost of GAC2 is c = (1d + 1e + 3v = 1.04), and the cost of GAC3 is c = (1d + 1v + 2i = 4.02). GAC2 is therefore the most likely explanation for the observed distributions in the TAC, given the cost assignments. Actually, GAC2 will remain optimal under a much wider range of cost assignments: as long as dispersals and extinctions cost more than vicariance and duplication events, the optimal solution will be the same (Sanmartín & Ronquist, 2002, Ronquist, 2003a). By using fast dynamic programming algorithms

(Ronquist, 1998b), it is possible to find the cost for any TAC–GAC combination. However, for problems with more than 10 areas, it is better to use heuristic algorithms similar to those used in ordinary parsimony analysis. These are implemented in TreeFitter 1.2 (Ronquist, 2003b).

AN EMPIRICAL EXAMPLE: *NOTHOFAGUS* BIOGEOGRAPHY

The widespread tree genus *Nothofagus* (Nothofagaceae), or southern beeches, is distributed in the Southern Hemisphere, with 35 species endemic to southern South America, Australia, New Zealand, New Caledonia, and New Guinea. The genus includes four subgenera: *Lophozonia, Brassospora, Fucospora,* and *Nothofagus;* all of them except *Nothofagus* are distributed in two or more southern continents but each species is endemic to one continent (Swenson et al., 2001). The biogeographic history of *Nothofagus* is considered as a classic example of the vicariance paradigm in the Southern Hemisphere (Manos, 1997; Swenson et al., 2001). The genus is thought to have originated in southern Gondwana, with the major lineages (subgenera) having already diverged in the Late Cretaceous prior to Gondwana's break up (Manos, 1997). The fragmentation of the supercontinent, starting in the Late Mesozoic, presumably divided the ancestral range of *Nothofagus* in successive vicariance events and gave rise to several endemic species in each of the southern continents. Because the biogeographic history of *Nothofagus* has been analyzed multiple times under different methods (Humphries, 1981; Humphries & Parenti, 1999; Linder & Crisp, 1995; Nelson & Ladiges, 1997, 2001; Swenson et al., 2001), it provides an excellent example to examine how parsimony-based tree fitting works.

Figure 6.2a shows the best area cladogram found by Brooks parsimony analysis, for a molecular phylogeny of 23 *Nothofagus* species and its associated distributions (Swenson et al., 2001; see Figure 6.2d) analyzed in PAUP 4.0 with a hypothetical outgroup of all zeros (Length = 23 steps). It shows southern South America (SSA) as the sister-group to an Australia–New Zealand clade, and these three areas are, in turn, the sister-group to the New Guinea–New Caledonia clade. TreeFitter finds not only this optimal area cladogram (cost = $15d + 6v + 1e + 1i = 3.21$), but also a second cladogram with the same cost (Figure 6.2b), in which SSA is the sister-group to the New Guinea–New Caledonia clade. Component analysis and subtree analysis will find these two cladograms and a third one, in which SSA is the sister-group to the rest of areas (SSA,((AUS,NZ),(NG,NC))), giving the consensus tree (SSA,(NG,NC)(AUS,NZ)) (Nelson & Ladiges, 2001; Swenson et al., 2001). In TreeFitter, however, the cost of the third cladogram is slightly higher than those in Figure 6.2a and 6.2b (c = $16d + 6v + 3e = 3.22$), so it is a less parsimonious solution.

None of these optimal cladograms, however, corresponds with the geological sequence of Gondwana break up represented in Figure 6.2c. When this geological area cladogram is mapped onto the phylogeny of *Nothofagus,* we obtain an event-based reconstruction (Figure 6.2d) that requires 16 duplication events, 3 vicariance events, 3 extinction events, and 3 dispersal events (c = $16d + 3v + 3e + 3i = 9.19$). This reconstruction is more costly than the optimal solutions in Figure 6.2a and 6.2b,

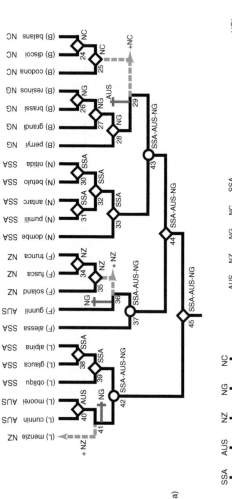
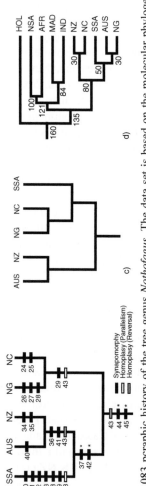

FIGURE 6.2 Biogeog38.083 pcraphic history of the tree genus *Nothofagus*. The data set is based on the molecular phylogeny of 23 species of *Nothofagus* (Swenson et al., 2001) representing all subgenera (L = *Lophozonia*, F = *Fuscospora*, N = *Nothofagus*, and B = *Brassospora*). (a) General area cladogram for this dataset found by Brooks parsimony analysis. (b) A second, alternative area cladogram found by TreeFitter. (c) Geological area cladogram representing the sequential break up of Gondwana; numbers indicate time of vicariance in mya (modif ed from Sanmartin & Ronquist, 2004). (d) Event-based reconstruction of *Nothofagus* biogeography showing the geological area cladogram mapped onto the phylogeny of *Nothofagus*, with symbols as in Figure 6.1. Abbreviations: AFR (Africa), AUS (Australia), HOL (Holarctic), IND (India), MAD (Madagascar), NC (New Caledonia), NG (New Guinea), NSA (Northern South America), NZ (New Zealand), SSA (Southern South America).

but represents the extent to which the biogeographic history of *Nothofagus* departs from the geological sequence of Gondwana break up (i.e., from a pure vicariance scenario). It postulates that the ancestor of *Nothofagus* was originally present in SSA, AUS, and NG at a time when these landmasses were still part of Gondwana. Subsequent duplication (speciation within the ancestral area) gave rise to subgenera *Fucospora* and *Lophozonia* and the ancestor of *Nothofagus* and *Brassospora*. Each of these subgenera is postulated to have undergone later vicariance between SSA / (AUS-NG), followed by extinction in NG (*Fucospora*, *Lophozonia*) or AUS (*Brassospora*). This was followed by extensive duplication within each continent and later dispersal from AUS to NZ in *Lophozonia* and *Fucospora*, or from NG to NC in *Brassospora*. Therefore, TreeFitter proposes a biogeographic scenario that combines classic Gondwana vicariance with subsequent dispersal events superimposed onto the original vicariant pattern. This is likely to be the case in many southern groups. Recently, Knapp *et al.* (2005), using DNA sequence data calibrated with a Bayesian relaxed molecular clock, concluded that the presence of *Lophozonia* and *Fucospora* in New Zealand is the result of recent dispersal events and not of tectonic vicariance. The TreeFitter reconstruction, however, is not that clear for subgenera *Nothofagus* and *Brassospora*. According to the fossil record, both subgenera were originally present in New Zealand and the other landmasses, and their present absence is the result of extinction (Manos, 1997; Swenson et al., 2001). Therefore, the event-based reconstruction in Figure 6.2a may be considered as the best reconstruction of the biogeographic history of *Nothofagus* based on *extant* taxa.

Compare the TreeFitter reconstruction with the BPA reconstruction in Figure 6.2a. The BPA reconstruction is based on the optimal area cladogram, not on the geological cladogram, but the most important difference is that the biogeographic history of *Nothofagus* can be read directly from the TreeFitter reconstruction (Figure 6.2d), whereas the BPA reconstruction requires careful *a posteriori* interpretation. The TreeFitter reconstruction directly specifies the direction of the implied dispersal events and the geographical location and relative timing of the vicariance, duplication, and extinction events (Figure 6.2d). By contrast, the BPA reconstruction can involve multiple, equally parsimonious interpretations. For example, to explain the presence of the ancestor of subgenera *Brassospora* and *Nothofagus* ("ancestral species 43") in areas SSA, NG, and NC, which do not form a clade in the GAC, the BPA reconstruction indicates that species 43 was originally present in all the areas but went subsequently extinct in areas AUS-NZ, a "reversal" (Figure 6.2b). An alternative, equally parsimonious reconstruction is a parallel occurrence of ancestral species 43 in areas NG-NC and in area SSA (i.e., a "convergence" or possible dispersal" event, Figure 6.2b). This example represents one of the problems of BPA. Because biogeographic events are not directly integrated into the analysis like in TreeFitter, any reconstruction that differs from the simple vicariance pattern (i.e., any case of incongruence or "homoplasy") involves multiple and alternative interpretations. Brooks (1990) suggested a modification of BPA to solve this problem, in which areas that appear to have a composite history in relation to the taxa analyzed are duplicated in the analysis ("secondary BPA", Brooks et al., 2001). However, this duplicating technique probably further complicates more than clarifies the interpretation of the resulting area cladograms.

Allegedly, component analysis and subtree analysis focus exclusively on searching for general patterns of area relationships ("area history"), without making any assumptions about the evolutionary processes involved (Ebach & Humphries, 2002, Ebach et al., 2003). However, in practice, processes such as duplications and extinctions are still invoked when interpreting incongruence between the individual taxon area cladograms and the general biogeographic pattern (e.g., Parenti & Humphries, 2004, Figure 1). Most biogeographers, in finding the general area cladogram in Figure 6.2a and 6.2b, would not stop there but try to explain any incongruities between the GAC and the individual biogeographic patterns. Subtree and component analyses (and reconciliation, Page, 1994) will explain the incongruities as cases of "paralogy", the result of duplication and extinction events ("inclusive OR-ing", Page, 1990), but they will not consider dispersal as a possible explanation. However, if all nodes are accounted for, in some cases an explanation that admits dispersal can be more likely (less costly) than one that does not, such as in the *Nothofagus* example. The success of pattern-based methods depends on the relation between the details of the method and the nature of the processes being inferred (Ronquist, 2003a). For example, component analysis will work well when the pattern analyzed has been caused by duplication and extinction events. Although dispersal can be incorporated into tree reconciliation through *ad hoc* procedures (i.e., pruning from the analysis the taxon suspected to have dispersed, Page, 1994, Swenson et al., 2001), dispersal events are not directly integrated into the reconstruction as in tree fitting or in maximum vicariance.

AREA BIOGEOGRAPHY: SOUTHERN HEMISPHERE BIOGEOGRAPHIC PATTERNS

Probably more interesting than inferring a general pattern of area relationships from a single group like *Nothofagus* is to infer a general biogeographic pattern from a set of different groups of organisms inhabiting the same region. In the largest biogeographic study attempted so far, Sanmartín & Ronquist (2004) used parsimony-based tree fitting to analyze the biogeographic history of the Southern Hemisphere based on the published phylogenies of 19 plant and 54 non-marine animal taxa (1393 terminals). The Southern Hemisphere has traditionally been considered as the model example of the vicariance scenario, but recent molecular and paleontological studies (Cooper et al., 2001; Pole, 2001) indicate that dispersal has played a bigger role in forming southern distributions than previously thought. To what extent do southern biogeographic patterns conform to the geological break up of Gondwana? Sanmartín & Ronquist (2004) found that animal patterns in general appear to have been generated by geological vicariance. The best GAC for the animal data set showed a sister-group relationship between Australia and South America, with New Zealand as sister-area (Figure 6.3a), in congruence with their break-up sequence from Gondwana (Figure 6.2c). By contrast, the plant data set supported a sister-group relationship between Australia and New Zealand, with South America as their sister-area (Figure 6.3b). This pattern does not conform to the geological scenario, but could be explained by trans-oceanic dispersal between Australia and New Zealand after the break up of land connections (Sanmartín & Ronquist, 2004).

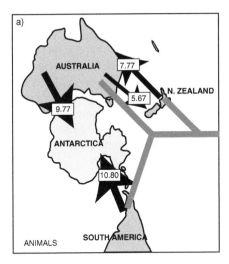
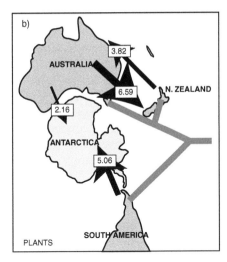

FIGURE 6.3 Example of parsimony-based tree fitting in area biogeography. (a) Polar view of the Southern Hemisphere in the Late Cretaceous showing the frequency of dispersal events among southern South America, Australia, and New Zealand inferred with parsimony-based tree fitting methods from a data set of 54 animal phylogenies (Sanmartín & Ronquist, 2004). Superimposed onto the figure is the general area cladogram for this data set. (b) Corresponding frequencies of dispersal events and the general area cladogram for the plant data set based on 19 phylogenies. Modified from Sanmartín & Ronquist (2004).

Tree-fitting methods can also be used to infer *dispersal patterns* from the phylogeny, providing these events are rare compared with the number of phylogenetically constrained events such as vicariance or duplication. For instance, Sanmartín & Ronquist (2004) searched for general patterns of dispersal (i.e., affecting multiple groups) in their phylogenies that could not be explained by the vicariance events predicted by a geologically based area cladogram. When animal phylogenies were fitted onto the geological cladogram in Figure 6.2c, trans-Antarctic dispersals were revealed as the dominant dispersal pattern. These are faunal links between Australia and South America that could not be explained by vicariance but probably reflec the long period of connection between these two continents, i.e., land dispersal (Sanmartín & Ronquist, 2004). In contrast, trans-Tasman dispersal between Australia and New Zealand was the dominant dispersal pattern in the plant data set. These two landmasses became separated 80 Mya ago, but they are still relatively close (2000 km). Also, dispersal from Australia to New Zealand is favored by westerly wind and ocean currents, the West Wind Drift and the Antarctic Circumpolar Current, which became established after the opening of the Drake Passage between South America and Antarctica in the Oligocene, 30 Mya ago. Pole (2001) argues that most of New Zealand's present flora came from Australia by long-distance dispersal driven by the West Wind Drift, and there are numerous historical records of birds and insects wind-blown across the Tasman Sea from Australia to New Zealand.

Thus, the most important conclusion from Sanmartín & Ronquist's (2004) analysis is that dispersal can generate congruent biogeographic patterns among multiple organisms in the same way as vicariance, if consistent in direction, source, and target areas (e.g., the West Wind Drift). This was a significant departure from the cladistic biogeographic notion that dispersal is a rare and stochastic event, unable to create congruent distribution patterns. Because dispersal is usually more stochastic in time than vicariance (McDowall, 2004), one way to distinguish between vicariance-induced hierarchical patterns and those generated by dispersal would be to correlate some estimate of divergence times between lineages with the time of the presumed geological vicariance event.

DISPERSAL-VICARIANCE ANALYSIS

Dispersal-vicariance analysis (DIVA) is an event-based method for reconstructing the biogeographic history of a group of organisms in the absence of a general hypothesis of relationships (Ronquist, 1997). Given an organism phylogeny and its associated geographic distributions, DIVA can reconstruct the optimal distributions at ancestral nodes and the biogeographic events that created those distributions. "Allopatric speciation associated with geographical vicariance" (Ronquist, 1997) is assumed as the null model, but at the same time DIVA accepts the potential contribution of extinction and dispersal in shaping the current distribution patterns.

Unlike parsimony-based tree fitting, where each event receives a positive cost, vicariance and duplication events have a cost of zero in DIVA, whereas dispersal and extinction events cost one per area added or deleted. Duplication implies no change in the distribution (as in ordinary parsimony analysis), whereas vicariance is the null model in DIVA (Ronquist, 1997). Nevertheless, all four biogeographic processes (vicariance, duplication, dispersal, and extinction) are considered as possible explanations for ancestral distributions in the DIVA analysis, as required by an event-based method.

Probably, the most important difference with TreeFitter is that DIVA does not assume that there is a single hierarchical pattern of area relationships (Ronquist, 1997). These are not enforced to conform to a general area cladogram as in TreeFitter, but are free to vary along the reconstruction. This is particularly appropriate when inferring biogeographic patterns in regions with a reticulate geological history, such as the Northern Hemisphere (Sanmartín et al., 2001) or the Mediterranean area (Sanmartín, 2003).

Dispersal-vicariance analysis works in a similar way as a character optimization method such as Fitch parsimony analysis. Distributions (characters) are optimized along the phylogeny to produce a reconstruction that minimizes the number of changes (dispersal and extinctions) in the ancestral distributions required for explaining the observed distributions in the terminals. However, Fitch parsimony analysis enforces ancestors to be monomorphic (restricted to single areas); only terminal taxa can be widespread (distributed in more than one area). In biogeographic terms, this means that only dispersal and duplication are accepted as explanations for the observed patterns. In contrast, DIVA allows both ancestors and terminals to be widespread (i.e., occurring in more than one area). Moreover, because (allopatric) speciation by vicariance is the null model in DIVA, vicariance and range division would always be the preferred explanation if ancestors are widespread (Ronquist, 1997).

Comparison with TreeFitter

Some simple examples (Figure 6.4) will illustrate the differences between dispersal-vicariance analysis and parsimony-based tree fitting. Consider a taxon-area cladogram with the two most basal lineages distributed in the same area (A). Dispersal-vicariance analysis (DIVA) explains this distribution by postulating vicariance of a widespread ancestor followed by secondary dispersal into A after the first vicariance to explain the presence of ancestor of species 2–5 in A. The cost of this reconstruction is $c = (4v + 1i)$. A true character optimization method such as Fitch parsimony analysis, which has duplication as the null model, would explain this distribution as two basal duplications in area A followed by consecutive dispersals (change of state) from A to B, B to C, and C to D. Because DIVA assumes vicariance as the null model, and dispersal costs one per area added to the distribution (+ B, C, D), the vicariance explanation is the preferred one here.

Figure 6.4b shows the TreeFitter reconstruction of the same data set illustrated as a trackogram, with the organism phylogeny fitted on top of the best area cladogram. It postulates a basal duplication in the ancestral area [ABCD], followed by extinction of the first lineage in [BCD] and vicariant division of the second lineage ($c = 1d + 3v + 1e$). DIVA would never find this reconstruction, because duplications can only occur within single areas in DIVA (cost zero). Duplications within widespread distributions such as (ABCD) are not allowed or would be equivalent to the number of secondary dispersals needed to explain that the two initially allopatric descendants (species 1 and the ancestor of 2–5) came to occupy the same areas (Ronquist, 1997). TreeFitter, however, does allow duplication within widespread distributions, but only if the areas forming the widespread distribution are postulated to have formed a contiguous region in the past according to the GAC, for example [ABCD] = G. In fact, TreeFitter (and maximum vicariance, Ronquist, 2003a) will only admit ancestral widespread distributions that form an ancestral area in the GAC. This is equivalent to forcing ancestral distributions to conform to a hierarchical vicariant pattern; vicariance events that do not agree with the GAC are simply not allowed (Sanmartín & Ronquist, 2002; Ronquist, 2003a).

Consider now the same TAC but with the distribution pattern inverted (Figure 6.4c–d). Both the DIVA and TreeFitter reconstructions suggest duplication in A, followed by dispersal into area B. The only difference is that the DIVA reconstruction (Figure 6.4c) requires one more vicariance event ($c = 1d + 3v + 1i$) than the TreeFitter reconstruction ($c = 1d + 2v + 1i$, Figure 6.4d). This is because dispersal events are always associated to speciation in tree-fitting methods (Page, 1995; Ronquist, 2003a). This means that the dispersal event will be mapped onto the node ancestral to the dispersal (the node of the ancestor of 4 and 5 in Figure 6.4d), so this node cannot be a vicariance event in the reconstruction. In DIVA, however, a dispersal event is always followed by a vicariance event (ancestral widespread distributions are divided at each speciation node, Ronquist, 1997), so the speciation event is associated to the node immediately after the dispersal. In Figure 6.4c, dispersal to B in the branch leading to the ancestor of species 4 and 5, is followed by a vicariance event between A and B.

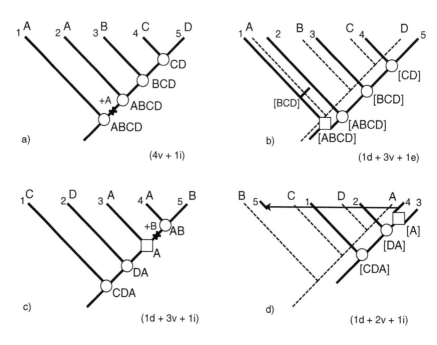

FIGURE 6.4 Some simple examples illustrating the differences between dispersal-vicariance analysis and parsimony-based tree fitting. (a–b) Taxon-area cladogram with the two most basal lineages distributed in the same area: (a) Dispersal-vicariance (DIVA) reconstruction (cost = 4v + 1i); (b) TreeFitter (TF) reconstruction of the same data set mapped onto the best area cladogram (A,(B,(C,D)) (c = 1d + 3v + 1e). DIVA will never find this reconstruction, because it does not allow duplication of widespread ancestors; this is allowed in TF but only if the widespread distribution forms an ancestral area in the area cladogram (indicated by brackets). Notice that both methods allow ancestors to be widespread; a true character optimization method such as Fitch parsimony analysis would explain this distribution as basal duplication in A followed by three dispersal events into B, C, and D. (c–d) The same TAC but with the distributions inverted. Both the DIVA and TF reconstructions suggest duplication in A followed by dispersal into area B, but TF (Figure 6.4d) requires one less vicariance event (c = 1d + 2v + 1i) than DIVA (c = 1d + 3v +1i, Figure 6.4c), because dispersal events are associated with speciation in parsimony-based tree fitting. (e–f) TAC with a redundant area (A), indicating conflicting area relationships: (e) DIVA postulates a dispersal event to explain the redundancy in A (c = 4v + i); (f) In the TF reconstruction, redundant distributions are treated in the same way as the other areas and resolved according to the most parsimonious area cladogram; in this case (B,(A,(C,D)) (c = 3v + 1i). (g–h) Treatment of widespread taxa in event-based methods. (g) DIVA explains this distribution as a terminal dispersal from B to A ("post-speciation dispersal in species 1), plus secondary dispersal to A by the ancestor of 3-5 (c = 4v + 1i + 1ti); (h) TF reconstruction under the *recent* option of dealing with widespread taxa (see text) also explains the widespread distribution as the result of recent dispersal; only A or B can be the true ancestral area, and the cost of each of them being ancestral will depend on the GAC they are fitted into. In this case, the optimal GAC would be (B,(A,(C,D))) because it only needs to assume an extra extinction and one duplication event to explain the observed distributions (c = 1d + 3v + 1e + 1ti). See text for further explanation.

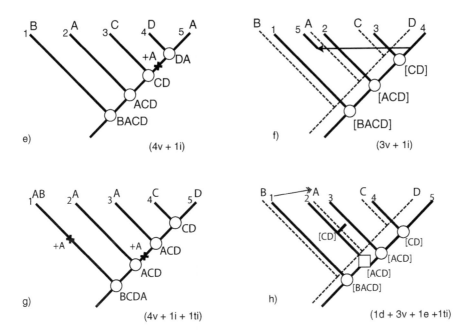

FIGURE 6.4 (Continued)

Figure 6.4e–f illustrates the treatment of redundant areas in DIVA and TreeFitter. *Redundant areas* are areas that harbor more than one taxon (e.g., area A in Figure 6.4a–f). They represent a source of ambiguity when inferring the general area cladogram, because they can occupy more than one position in the cladogram, and thus indicate different area relationships, for example, (B,(A,(C,D))) or (B,(C,(D,A))) in Figure 6.4e. These problems are resolved in pattern-based methods by using Assumptions 0, 1, and 2, which differ in the restrictions they apply to the set of possible solutions (topologies) accepted for the GAC (Nelson & Platnick, 1981; Van Veller et al., 2000, Ebach & Humphries, 2002). In parsimony-based tree fitting, redundant areas do not represent a particular problem in the GAC inference. They are treated in the same way as the rest of areas and are resolved according to the most parsimonious area cladogram. In this case, the best GAC for the observed distributions is (B,(A,(C,D))), because it only requires vicariance events and one dispersal to area A by the ancestor of 4–5 to explain the redundant distribution (c = 3v + 1i = 2.03). The alternative pattern (B,(C,(D,A))) would require three extinction events in B, C, and D (cost = 1d + 2v + 3e = 3.03). DIVA shows the same reconstruction as TreeFitter except for the "extra" vicariance event associated to the dispersal (c = 4v + 1i).

TREATMENT OF WIDESPREAD TAXA

Widespread taxa (terminal taxa distributed in more than one area) also pose a problem in biogeographic reconstruction because they introduce ambiguity in the data set, and solutions to this problem have been extensively discussed in "pattern-based" literature (Ebach et al., 2003; Van Veller et al., 2003, and related papers). I discuss here the treatment of widespread taxa within the event-based approach (see Sanmartín & Ronquist, 2002, Ronquist, 2003a, for further explanation). Consider a phylogeny with a widespread taxon (species 1). DIVA (Figure 6.4g) explains this distribution by assuming a "terminal" dispersal event into A by species 1 ("post-speciation dispersal"), followed by a second, "ancestral", dispersal to A along the branch leading to the ancestor of 3–5; cost = (4v + 1i + 1ti). The tree-fitting solution is more complicated because it needs to deal with the assumption of "one taxon per area" associated to the constraint of area relationships being hierarchical (i.e., conforming to a GAC). Sanmartín & Ronquist (2002) proposed three different options to deal with widespread taxa within the event-based approach. In the *recent* option, the widespread distribution is assumed to be the result of recent dispersal, so only one of the areas in the distribution can be ancestral; in the *ancient* option, the widespread distribution is assumed to be of ancestral origin and explained by vicariance and extinction; in the *free* option, the widespread distribution is treated as an unresolved higher taxon, and any combination of events (e.g., dispersal, extinction, vicariance) is accepted to explain it. Empirical (Sanmartín et al., 2001) and simulation (Ronquist, 2003a) studies showed that the recent option is more powerful than the free or ancient options in the identification of phylogenetically constrained biogeographic patterns. The reason for this is that the recent option disregards the widespread distribution as the result of recent dispersal. These dispersal events are not counted when calculating the cost of the optimal GAC(s), because they are also present across all possible GACs and random data sets. However, this actually results in the vicariance events which are not allowed within the widespread terminals being pushed further down onto the ancestral nodes in the phylogeny. For example, in Figure 6.4h, the recent option considers the widespread distribution in species 1 as the result of a terminal dispersal; only A and B can be ancestral areas. So the best area cladogram for this data set is (B,(A,(C,D)), because it will only need one extra extinction and one duplication event to explain the observed distribution pattern. The ancient option, by contrast, would infer ((A,B),(C,D)) as the best GAC, because it forces widespread terminal distributions to be interpreted as supporting area relationships. The vicariance event between A and B is more basal in the GAC obtained under the recent option than in the GAC obtained with the ancient option. This basal vicariance pattern is less likely to happen in the random data sets because the distributions are shuffled in the phylogeny, so it is easier to separate phylogenetically constrained patterns from random patterns in the recent option than in the ancient option. In practice, this means that, in building area relationships, the recent option effectively disregards the widespread distribution in favor of the endemics.

WHEN TO USE DIVA?

DIVA is the appropriate method when we want to infer the biogeographic history of individual or multiple lineages in the absence of a general area cladogram, for example, when (a) the region analyzed shows a "reticulate" biogeographic history such as the Holarctic, or (b) the group studied is much younger than the areas it occurs so it could not have been affected by any of the vicariance events that split those areas. In this case, the group distribution is better explained by dispersal to new areas followed by allopatric speciation rather than by real tectonic vicariance. When DIVA is not indicated? When there is a strong underlying vicariance pattern in the biogeographic distributions such as in the biogeographic history of the Southern Hemisphere (Sanmartín & Ronquist, 2004). In that case, DIVA loses power of resolution in comparison with an event-based tree fitting method such as TreeFitter. For example, a DIVA reconstruction of the biogeographic history of *Nothofagus* (not shown) will find multiple alternative distributions for most of the ancestral nodes (e.g., SSA-AUS-NZ / SSA-AUS / SSA-NZ for ancestor 42 in Figure 6.2a). Some of these reconstructions imply vicariance events that are incompatible with each other (SSA-AUS / SSA-NZ), or that do not conform to the geological vicariance scenario (SSA-NZ).

AN EMPIRICAL EXAMPLE: HOLARCTIC BIOGEOGRAPHY

Consider the organism phylogeny shown in Figure 6.5a, with 14 species occurring in the four Northern Hemisphere landmasses: Eastern Nearctic (EN), Western Nearctic (WN), Eastern Palearctic (EP), and Western Palearctic (WP). When BPA was used to analyze the biogeographic history of this group, it found a general area cladogram (Figure 6.5b) indicating the current continental configuration: a sister-group relationship between the Nearctic (EN-WN) and Palearctic (EP-WP) regions, followed by east-west vicariance. TreeFitter will find another area cladogram showing a sister-group relationship between EN-WP and WN-EP, followed by east-west vicariance (Figure 6.5c). According to paleogeographic reconstructions (Smith et al., 1994), during the Cretaceous, the Holarctic landmasses were joined in a different configuration to the one we observe today: eastern North America and the western Palearctic were connected, forming the paleocontinent of *Euramerica* (EN-WP), whereas western North America and the Eastern Palearctic formed the paleocontinent of *Asiamerica* (WN-EP). Therefore, BPA and TreeFitter reflect the continental configuration at two different time periods: present (BPA) and ancient (TF). One explanation for this difference is the different way BPA and TreeFitter deal with widespread taxa. In BPA, widespread taxa are considered as *bona-fide* evidence of area relationships (Van Veller et al., 2003), so the BPA cladogram reflects the fact that several terminals and near-terminal ancestors present widespread (two-area) continental distributions. In contrast, if the recent option is used, TreeFitter will disregard widespread taxa as the result of terminal dispersals (which are not counted in the reconstruction) and infer area relationships only from the deep phylogeny nodes, reflecting an older configuration (EN-WP/WN-EP).

However, when compared with paleogeographic reconstructions of the Holarctic region (Figure 6.5d), none of these area cladograms reflects the actual geological history. As mentioned above, Holarctic biogeography conforms to a "reticulate" scenario, in which continents joined, split, and became joined again through time, resulting in the repeated merging and fragmenting of biotas. Before time t1 (Cretaceous), we

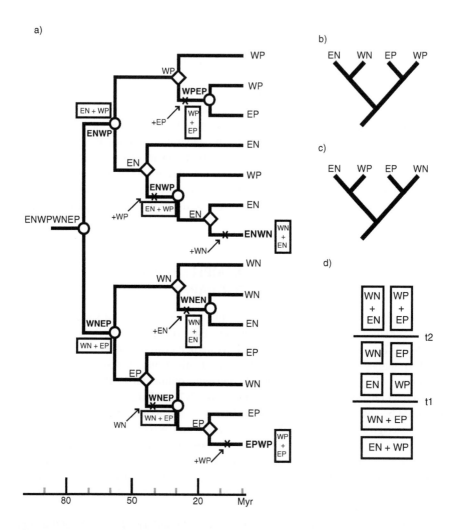

FIGURE 6.5 (a) Dispersal-vicariance reconstruction of the biogeographic history of a group of organisms distributed in the Holarctic landmasses (EN = Eastern Nearctic, WN = Western Palearctic, EP = Eastern Palearctic, WP = Western Palearctic); (b) General area cladogram for this data set found by BPA; (c) General area cladogram found by TreeFitter; (d) Schematic representation of the Cenozoic geological history of the Holarctic (Smith et al., 1994), exemplifying a reticulate biogeographic scenario which cannot be represented into a branching area cladogram.

find two paleocontinents, Asiamerica and Euramerica, with their separate biotas. In time t1, the opening of the Atlantic Ocean and the Beringian Strait split the landmasses, with the resulting vicariance. In time t2 (Late Tertiary), the continents became joined again, but in the current continental configuration (Eurasia and North America), followed by extensive dispersal within Nearctic and Palearctic biotas. Dispersal-vicariance analysis (Figure 6.5a) is able to recover this complex, reticulate biogeographic pattern, because it does not force area relationships to conform to a hierarchical area cladogram as BPA or TreeFitter do. Areas are allowed to combine in any given configuration in the reconstruction, so vicariance events inferred at some ancestral nodes (EN-WP/WN-EP) can be incompatible with vicariance events inferred further down in the phylogeny (EN-WN/EP-WP). The TreeFitter and BPA cladograms are able to recover the biogeographic pattern dominant in a particular time slice (t1 and t2, respectively), but DIVA allows us to recover the complete biogeographic scenario. If we are able to associate the nodes in the Holarctic example with some time estimate (Figure 6.5a), we could observe that the inferred vicariance events and their associated widespread distributions correspond to periods when the Holarctic landmasses were connected. For example, the firs distributions involving EN-WP / WN-EP would correspond to the Euramerica–Asiamerica connections (>80 Mya, Figure 6.5a), whereas the later paleocontinental distributions could correspond to the Trans-Atlantic and the Beringian Bridges that connected North America and Eurasia from the Early-Mid Tertiary onward (Sanmartín et al., 2001).

AREA BIOGEOGRAPHY: HOLARCTIC BIOGEOGRAPHIC PATTERNS

Unlike TreeFitter, dispersal-vicariance analysis is primarily a method to reconstruct the biogeographic history of individual groups. However, DIVA can also be used to find general patterns of dispersal and vicariance affecting many organisms simultaneously. This is done by analyzing each group separately and then calculating the frequencies of particular dispersal-vicariance events among the studied groups. For example, Sanmartín *et al.* (2001) used dispersal-vicariance analysis to search for common patterns of dispersal, vicariance, and diversification in the Holarctic based on a large data set of 57 animal phylogenies (770 species). By associating each node in the phylogeny with a time estimate, they could identify former land connections (e.g., the Trans-Atlantic Bridge), confirming paleogeographic reconstructions. Moreover, by comparing the relative frequency of dispersal-vicariance events with the frequencies expected by chance if distributions were not phylogenetically conserved, (Sanmartín et al., 2001) were able to test specific hypotheses concerning Northern Hemisphere biogeography. For example, their analysis revealed that the North Atlantic was the main dispersal route for Holarctic faunas during the Early Tertiary, whereas faunal migration over the Beringian Bridge was probably controlled by the prevailing climatic conditions, so its importance as a dispersal route changed considerably over time. Comparison of phylogeny-based estimates of species richness at different time periods also suggested that the present differences in species richness among Holarctic infraregions (i.e., higher faunal diversity in Asia) is probably the result of differential diversification during the Tertiary, rather than the effect of Pleistocene climatic changes (Latham & Ricklefs, 1993). Recently, Donoghue & Smith (2004) repeated Sanmartín *et al.*'s study using DIVA to analyze 66 phylogenies of plant taxa with mainly Holarctic distributions. Their study confirmed a reticulate Holarctic scenario but showed that plant patterns were, in general, younger than those in animals, implying more recent dispersal (connection) events (although they did not test for the significance of patterns). This agreed with Sanmartín & Ronquist's (2004) study on the Southern Hemisphere, which shows animals patterns as largely congruent with Gondwana's breakup, whereas plant patterns were better explained by recent trans-oceanic dispersal. If these patterns are not the result of artifacts in the analysis but reflect major biogeographic differences between plants and animals, this could have significant implications for community assembly and coevolution (Donoghue & Smith, 2004). It would suggest that plants are more efficient than animals in colonizing new areas, perhaps due to the greater capacity of plant seeds to disperse (higher vagility), whereas animals show higher resilience (i.e., the ability to cope with changing climates and extinction) and thus reflect older patterns in their phylogenies.

ACKNOWLEDGEMENTS

I am grateful to Malte Ebach for inviting me to participate in the Symposium "What is Biogeography?" held at the 2005 Systematics Association Biennial in Cardiff, on which I presented this paper. I also thank the students at the Systematic Biology Course of Uppsala University, and an anonymous reviewer, for input and ideas that helped me to clarify this paper. This work was supported by the Swedish Science Research Council (Grant 621-2003-456).

REFERENCES

Altaba, C. (1998). Testing vicariance: melanopsid snails and Neogene tectonics in the western Mediterranean. *Journal of Biogeography,* 25, 541–551.

Andersson, L. (1996). An ontological dilemma: epistemology and methodology of historical biogeography. *Journal of Biogeography,* 23, 269–277.

Baum, D.A., Small, R.L. & Wendel, J.F. (1998). Biogeography and floral evolution of baobabs (*Adansonia,* Bombaceae) as inferred from multiple data sets. *Systematic Biology,* 47, 181–207.

Brooks, D.R. (1985). Historical ecology: a new approach to studying the evolution of ecological associations. *Annals of the Missouri Botanical Garden,* 72, 660–680.

Brooks, D.R. (1990). Parsimony analysis in historical biogeography and coevolution: methodological and theoretical update. *Systematic Zoology,* 39, 14–30.

Brooks, D.R. & McLennan, D.A. (2001). A comparison of a discovery-based and an event-based method of historical biogeography. *Journal of Biogeography,* 28, 757–767.

Brooks, D.R., Van Veller, M. & McLennan, D.A. (2001). How to do BPA, really. *Journal of Biogeography,* 28, 343–358.

Brundin, L. (1966). Transantarctic relationships and their significance, as evidenced by chironomid midges with a monograph of the subfamilies Podonominae and Aphroteninae and the austral Heptagynae. *Kungliga Svenska Vetenskapsakademia Handlingar,* 11, 1–472.

Buckley, T.R., Arensburger, P., Simon, C. & Chambers, G.K. (2002). Combined data, Bayesian phylogenetics, and the origin of the New Zealand Cicada genera. *Systematic Biology,* 51, 4–18.

Cook, L.G. & Crisp, M.D. (2005). Directional asymmetry of long-distance dispersal and colonization could mislead reconstructions of biogeography. *Journal of Biogeography,* 32, 741–754.

Cooper, A.C., Lalueza-Fox, C., Anderson, S., Rambaut, A., Austin, J. & Ward, R. (2001). Complete mitochondrial genome sequences of two extinct moas clarify ratite evolution. *Nature* 409, 704–707.

Craw, R. (1982). Phylogenetics, areas, geology and the biogeography of Croizat: a radical view. *Systematic Zoology*, 31, 304–316.

Crisp, M.D. & Cook, L.G. (2004). Do early branching lineages signify ancestral traits? *Trends in Ecology and Evolution*, 20, 122–128.

Croizat, L. (1964). *Space, Time, Form: The Biogeographical Synthesis.* Privately published. Caracas, Venezuela.

Croizat, L., Nelson, G. & Rosen, D.E. (1974). Centres of origin and related concepts. *Systematic Zoology*, 23, 265–287.

Cunningham, C.W. & Collins, T.M. (1994). Developing model systems for molecular biogeography: vicariance and interchange in marine invertebrates. In *Molecular Ecology and Evolution, Approaches and Application* (ed. by Schierwater, B. Streit, B., Wagner, G.P. & DeSalle, R.), pp. 405–433. Birkhauser Verlag Basel, Switzerland.

Darlington P.J. (1957). *Zoogeography: The Geographical Distribution of Animals.* John Wiley & Sons, New York.

Donoghue, M.J. & Moore, B.R. (2003). Toward an integrative historical biogeography. *Integrative Comparative Biology*, 43, 261–270.

Donoghue, M.J. & Smith, S. (2004). Patterns in the assembly of temperate forests around the Northern Hemisphere. *Philosophical Transactions of the Royal Society of London, Serie B*, 359, 1633–1644.

Ebach, M.C. & Humphries, C.J. (2002). Cladistic biogeography and the art of discovery. *Journal of Biogeography*, 20, 427–444.

Ebach, M., Humphries, C.J. & Williams, D.M. (2003). Phylogenetic biogeography deconstructed. *Journal of Biogeography*, 30, 1258–1296.

Enghoff, H. (1995). Historical biogeography of the Holarctic: area relationships, ancestral areas, and dispersal of non-marine animals. *Cladistics*, 11, 223–263.

Heads, M. (1999). Vicariance biogeography and terrane tectonics in the South Pacific: analysis of the genus *Abrotanella* (Compositae). *Biological Journal of the Linnean Society*, 67, 391–432.

Hennig, W. (1966). *Phylogenetic Systematics.* University of Illinois Press, Urbana.

Humphries, C.J. (1981). Biogeographical methods and the southern beeches (Fagaceae: *Nothofagus*). In *Advances in Cladistics. Proceedings of the First Meeting of the Willi Hennig Society* (ed. by. Funk, V.A & Brooks, D.R.), pp. 177–207. New York Botanical Garden, Bronx, NY.

Humphries, C.J. (2001). Vicariance biogeography. In *Encyclopedia of Biodiversity*, Vol. 5 (ed. by Levin, S.), pp. 767–779. Academic Press, San Diego, CA.

Humphries, C.J. & Parenti, L. (1986). *Cladistic Biogeography.* Clarendon Press, Oxford.

Humphries, C.J. & Parenti, L. (1999). *Cladistic Biogeography: Interpreting Patterns of Plant and Animal Distribution.* Oxford University Press, New York.

Knapp, M., Stöckler, K., Havell, D., Delsuc, F., Sebastiani, F. & Lockhart, P.J. (2005). Relaxed molecular clock provides evidence for long-distance dispersal of *Nothofagus* (southern beech). *PLOS Biology*, 3(1), e14.

Linder, H.P. & Crisp, M.D. (1995). *Nothofagus* and Pacific Biogeography. *Cladistics*, 11, 5–32.

Latham R.E & Ricklefs R.E. (1993). Continental comparisons of temperate-zone tree species diversity. In *Species Diversity in Ecological Communities* (ed. by Ricklefs, R.E. & Schluter, D.), pp. 294–314. The University Chicago Press, Chicago.

Manos, P.S. (1997). Systematics of *Nothofagus* (Nothofagaceae) based on rDNA spacer sequences (ITS): taxonomic congruence with morphology and plastid sequences. *American Journal of Botany,* 84, 1137–1155.

McDowall, R.M. (2004). What biogeography is: a place for process. *Journal of Biogeography,* 31, 345–351.

Morrone, J.J. & Crisci, J.V. (1995). Historical biogeography: introduction to methods. *Annual Review of Ecology and Systematics,* 26, 373–401.

Muñoz, J., Felicísimo, A.M., Cabezas, F., Burgaz, A.R. & Martínez, I. (2004). Wind as a long-distance dispersal vehicle in the Southern Hemisphere. *Science,* 304, 1144–1147.

Nelson, G. & Ladiges, P. (1996). Paralogy in cladistic biogeography and analysis of paralogy-free subtrees. *American Museum Novitates,* 3167, 1–58.

Nelson, G. & Ladiges, P. (1997). Subtree analysis, *Nothofagus* and Pacific biogeography. *Cladistics,* 13, 125–129.

Nelson, G. & Ladiges, P. (2001). Gondwana, vicariance biogeography and the New York School revisited. *Australian Journal of Botany,* 49, 389–409.

Nelson, G. & Platnick, N.I. (1981). *Systematics and Biogeography: Cladistics and Vicariance.* Columbia University Press, New York.

Page, R.D.M. (1990). Component analysis: A valiant failure? *Cladistics,* 6, 119–136.

Page, R.D.M. (1994). Maps between trees and cladistic analysis of historical associations among genes, organisms, and areas. *Systematic Biology,* 43, 58–77.

Page, R.D.M. (1995). Parallel phylogenies: Reconstructing the history of host-parasite assemblages. *Cladistics,* 10, 155–173.

Page, R.D.M. (2003). Introduction. In *Cospeciation* (ed. by Page, R.D.M.), pp. 1–22. Chicago University Press, Chicago.

Page, R.D.M. & Charleston, M.A. (1998). Trees within trees: phylogeny and historical associations. *Trends in Ecology and Evolution* 13, 356–359.

Parenti, L. & Humphries, C.J. (2004). Historical biogeography, the natural science. *Taxon,* 53, 899–903.

Platnick, N. & Nelson, G.J. (1978). A method of analysis for historical biogeography. *Systematic Zoology,* 27, 1–16.

Pole, M.S. (2001). Can long-distance dispersal be inferred from the New Zealand fossil record? *Australian Journal of Botany,* 49, 357–366.

Ronquist, F. (1994). Ancestral areas and Parsimony. *Systematic Biology,* 43, 267–274.

Ronquist, F. (1995). Reconstructing the history of host-parasite associations using generalized parsimony. *Cladistics,* 11, 73–89.

Ronquist F. (1996). DIVA, ver. 1.1. User's Manual. Computer program for MacOS and Win32. Available from http://www.ebc.uu.se/systzoo/research/diva/ diva.html.

Ronquist, F. (1997). Dispersal-vicariance analysis: a new biogeographic approach to the quantification of historical biogeography. *Systematic Biology,* 46, 195–203.

Ronquist, F. (1998a). Phylogenetic approaches in coevolution and biogeography. *Zoologica Scripta,* 26, 313–322.

Ronquist, F. (1998b). Three dimensional cost-matrix optimization and minimum cospeciation. *Cladistics* 14, 167–172.

Ronquist, F. (2003a). Parsimony analysis of coevolving species associations. In *Cospeciation* (ed. by Page, R.D.M.), pp. 22–64. Chicago University Press, Chicago.

Ronquist, F. (2003b). TreeFitter v. 1.2. Software available from http://www.ebc.uu.se/systzoo/research/treefitter/treefitt .html.

Ronquist, F. & Nylin, S. (1990). Process and pattern in the evolution of species associations. *Systematic Zoology,* 39, 323–344.

Sanmartín, I. (2003). Dispersal vs. vicariance in the Mediterranean: historical biogeography of the Palearctic Pachydeminae (Coleoptera, Scarabaeoidea). *Journal of Biogeography*, 30, 1883–1897.

Sanmartín, I., Enghoff, H. & Ronquist, F. (2001). Patterns of animal dispersal, vicariance and diversification in the Holarctic. *Biological. Journal of the Linnean Society*, 73, 345–390.

Sanmartín, I. & Ronquist, F. (2002). New solutions to old problems: widespread taxa, redundant distributions and missing areas in event-based biogeography. *Animal Biodiversity and Conservation*, 25, 75–93.

Sanmartín, I. & Ronquist, F. (2004). Southern hemisphere biogeography inferred by event-based models: plants versus animal patterns. *Systematic Biology*, 53, 216–243.

Seberg, O. (1988). Taxonomy, phylogeny and biogeography of the genus *Oreobolus* R. Br. (Cyperaceae), with comments on the biogeography of the South Pacific continents. *Botanical Journal of the Linnean Society*, 96, 119–195.

Simberloff, D., Heck, K.L., McCoy, E.D. & Conner, E.F. (1981). There have been no statistical tests of cladistic biogeographic hypotheses. In *Vicariance Biogeography: A Critique* (ed. by Nelson, G. & Rosen, D.E.), pp. 40–63. Columbia University Press, New York.

Smith A.G., Smith D.G. & Funnell B.M. (1994). *Atlas of Mesozoic and Cenozoic coastlines*. Cambridge University Press, Cambridge.

Swenson, U., Hill, R.S. & McLoughlin, S. (2001). Biogeography of *Nothofagus* supports the sequence of Gondwana break-up. *Taxon*, 50, 1–17.

Van Veller, M.G.P., Kornet, D.J. & Zandee, M. (2000). Methods in vicariance biogeography: assessment of the implementations of assumptions zero, 1, and 2. *Cladistics*, 16, 319–345.

Van Veller, M.G.P., Brooks, D.R. & Zandee, M. (2003). Cladistic and phylogenetic biogeography: the art and the science of discovery. *Journal of Biogeography*, 30, 319–329.

Vilhelmsen, L. (2004). The old wasp and the tree: fossils, phylogeny and biogeography in the Orussidae (Insecta, Hymenoptera). *Biological Journal of the Linnean Society*, 82, 139–160.

Voelker, G. (1999). Molecular evolutionary relationships in the avian genus *Anthus* (Pipits: Motacillidae). *Molecular Phylogenetics and Evolution*, 11, 84–94.

Waters, J.M., Dijkstra, L.H. & Wallis, G.P. (2000). Biogeography of a southern hemisphere freshwater fish: how important is marine dispersal? *Molecular Ecology*, 9, 1815–1821.

Wiley, E.O. (1988). Parsimony analysis and vicariance biogeography. *Systematic Zoology*, 37, 271–290.

Winkworth, R.C., Wagstaff, S.J., Glenny, D. & Lockhart, P.J. (2002). Plant dispersal N.E.W.S. from New Zealand. *Trends in Ecology and Evolution*, 17, 514–520.

7 Phylogeography in Historical Biogeography: Investigating the Biogeographic Histories of Populations, Species, and Young Biotas

Brett R. Riddle and David J. Hafner

ABSTRACT

Over the past two decades, phylogeography has become an increasingly popular approach to investigating the geography of genetic variation within and among populations, species, and groups of closely related species. Phylogeographic research is uniquely positioned between historical and ecological biogeography, but to date has not incorporated many of the fundamental concepts of the former and, therefore, is susceptible to criticism that it is not a legitimate method in area-based historical biogeography. Here, we review the similarities and differences between phylogeography and area-based historical biogeography; and review concerns regarding the differences. We then summarize one recent approach to reconciling differences that highlights the synergistic and reciprocal strengths of each approach at different stages in the analysis of historical structure within populations, species, and young biotas.

INTRODUCTION

In recent years, the number and variety of new approaches and methods in historical biogeography has grown steadily (summarized by Crisci et al., 2003; see also table 12.2 in Lomolino et al., 2006), but not without debate over the utility and validity of many of them (e.g., Brooks et al., 2001, 2004; Ebach, 2001; Van Veller & Brooks, 2001; Ebach & Humphries, 2002, 2003; Siddall & Perkins, 2003; Van Veller et al., 2003; Siddall, 2005). Much of this recent diversification of historical biogeography can be attributed to a desire to explore aspects of biotic histories that were not considered tractable under the original form of *vicariance* (or *cladistic*) *biogeography* that emerged in the 1970s (e.g., Nelson, 1974; Crisci et al., 2003). Heated discussions have focused on the range of processes that historical biogeographers should be investigating: e.g., restricted to associations between Earth and biotic histories (Humphries & Ebach, 2004) vs. a full range of vicariance and dispersal histories that underlay the geography of diversification (Brooks, 2004). Additionally, historical biogeography has been challenged on two fronts: first, for not being able to accomplish fundamental tasks such as deciphering the complexity (*pseudo-congruence*) often embedded within a single postulated vicariance event (Donoghue & Moore, 2003), and second, for not yet reaching its potential to contribute to more fundamental questions in ecology and evolution (Wiens & Donoghue, 2004).

One topic of recent debate concerns the role within historical biogeography of *phylogeography,* the extraordinarily popular approach (Figure 7.1a) that began in the late 1980s (Avise et al., 1987). Avise (2000) defined phylogeography as "a fiel of study concerned with the principles and processes governing the geographic distributions of genealogical lineages, especially those within and among closely related species." The status of phylogeography as a legitimate method in biogeography has been challenged (e.g., Humphries, 2000; Ebach & Humphries, 2002) and, in response, defended (Arbogast & Kenagy, 2001; Riddle & Hafner, 2004). Here, we attempt to clarify unique attributes that have arisen within a phylogeographic paradigm that we believe provide not only a legitimate but an increasingly important component to integrative approaches that provide investigators with the power to address the rich array of patterns and processes involved in the biogeographic histories of populations, species, and young biotas.

PHYLOGEOGRAPHY VS. HISTORICAL BIOGEOGRAPHY

While phylogeographic and historical biogeographic studies overlap in their connections to a variety of macroevolutionary topics (Figure 7.1b), they differ in a number of ways, the most obvious being that phylogeographic studies do not generally extend beyond a Neogene timeframe and typically target a Pleistocene timeframe (Figure 7.1c). As such, phylogeographic approaches have provided a basis for expanding basic questions in historical biogeography into the realms of Quaternary paleoecology and macroecology. Within this timeframe, processes bearing on diversification and distributional dynamics within and among species prominently feature paleoclimate of the Pleistocene (Hewitt, 2004). However, phylogeographic studies often reveal levels of divergence consistent with Pliocene or Miocene timeframes, which require a consideration of the very profound effects of Neogene geotectonics on current landscapes (e.g., Riddle et al., 2000; Mateos, 2005).

Phylogeography gained much of its current popularity by providing a linkage, via molecular population genetic and phylogenetic analyses, between those disciplines concentrating on population and species responses to relatively recent evolutionary processes or paleoclimatic events in Earth history with those that have traditionally been more concerned with the association between biological diversification and geological events in Earth history. This "microevolution/macroevolution"

a

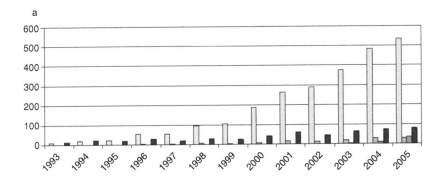

b

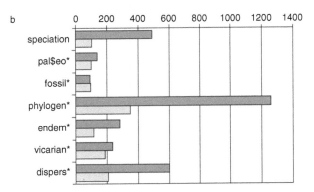

c

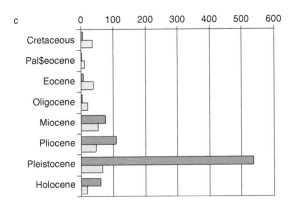

FIGURE 7.1 Several comparisons between phylogeography and historical biogeography based on title, abstract, and keyword searches on the ISI Web of Science, 1993–2005. (a) Total number of hits using, from left to right in each year, "phylogeograph*", "comparative phylogeograph*", "statistical phylogeograph*", or "historical biogeograph* not phylogeograph*". In graphs (b–e), searches were performed using either "phylogeograph* and [the term on the graph]" (dark grey bars) or "historical biogeograph* not phylogeograph* and [the term on the graph]" (light grey bars). (See text for discussion).

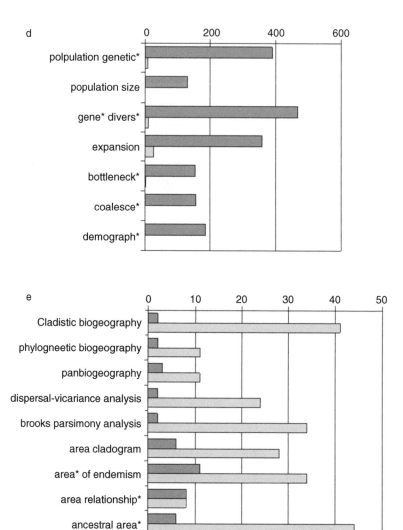

FIGURE 7.1 (Continued)

character should position phylogeographic-based research in a prominent role as historical biogeographers increasingly incorporate questions traditionally of great interest in ecology and evolutionary biology (e.g., Brooks, 2004; Wiens & Donoghue, 2004; Riddle, 2005). Several features, however, suggest a strong disconnect between phylogeography and much of the rest of historical biogeography. First, although phylogeography, since its inception, has been considered as a bridge between traditionally separate macroevolutionary and microevolutionary (Figure 7.1d) regimes (Avise, 2000), phylogeographic studies increasingly explore dispersal and population connectivity within ongoing or very recent ecological arenas (see any recent issue of the journal *Molecular Ecology* for examples). Within these studies, historical biogeographic explanations are included as but one alternative or one level in a spatio-temporal hierarchy of explanations. Increasing sophistication of coalescent-based analytical methods and capabilities to sample different (particularly rapidly evolving) partitions of an organism's genome (e.g., mitochondrial vs. nuclear) has driven much of phylogeography in the direction of population genetic analyses of very recent events in the history of a population or species and has led to the introduction of the concept of *statistical phylogeography* (Knowles, 2004; Templeton, 2004). For example, the goal of nested clade phylogeographic analysis (Templeton, 2004) is to differentiate between the influences of population structure (processes that determine gene fl w among extant populations) and population history (vicariance, long-distance dispersal, etc.) in determining current patterns of genetic diversity across the geographic range of a species.

A second indication of a disconnect between phylogeography and historical biogeography is the near lack in phylogeography of reference to an array of terms, concepts, and methods that are central to the rest of historical biogeography (Figure 7.1e). Despite the explosive popularity of phylogeographically based studies, very few incorporate the concept of an *area of endemism,* which is the fundamental unit of analysis in much of the rest of historical biogeography. A search on ISI Web of Science, 1993–2005, revealed a total of 2,495 studies with "phylogeograph*" in title, abstract, or keywords, yet a search for "phylogeograph* and area* of endemism" (Figure 7.1e) revealed only 11 papers, three of which referred to a single study. Likewise, "phylogeograph* and area cladogram" produced only six hits. Clearly, phylogeographic studies rarely are designed to incorporate the conceptual framework that predominated in historical biogeography over the past three decades. For this reason alone, it should not be surprising that researchers operating within a more traditional cladistic biogeographic framework have expressed a concern about the legitimacy of phylogeography within historical biogeography (Humphries, 2000; Ebach & Humphries, 2003; Humphries & Ebach, 2004).

Cladistic biogeographers have correctly criticized phylogeography for its failure to incorporate the most important constructs of late twentieth-century historical biogeography, including the frequent reliance in phylogeographic studies on single-taxon approaches to differentiate between dispersal and vicariance histories. Nevertheless, we believe that the criticism that phylogeography is weighted toward the production of ad hoc dispersal scenarios (and thus resurrects a failed paradigm in biogeography) has discounted two areas of growth in phylogeography: statistically rigorous population genetic methods for assessing population geographic structure and history, and the emerging popularity of comparative phylogeographic approaches.

Many of the methods incorporated into statistical phylogeography (Knowles, 2004; Templeton, 2004) are normally used to address hypotheses about the geographic structure and history of populations. These include post-Pleistocene range expansion of a single species out of one or more glacial-age refugia (Hewitt, 2004), the prevailing direction of gene fl w via dispersal between two or more geographically separated populations within a single species (Zheng et al., 2003), and the relative contributions of historical and extant population processes in structuring geographical population genetic architecture (Templeton et al., 1995). These approaches incorporate a rich theory based on population genetic principles (migration, drift, selection, effective population sizes) to differentiate, for example, between either long-term occupancy of a current and widespread geographic distribution, or recent range expansion from a previously smaller distribution. As such, they rely on a very different set of acceptance or rejection criteria to test historical hypotheses. In contrast, area-based methods in biogeography usually rely on examination of phylogenetic relationships across an *a priori* chosen set of areas of endemism, and many further arenas using discordance with a general area cladogram to suggest that the biogeographic history of taxa across a set of areas of endemism is at odds with an underlying vicariant history.

However, as the timeframe for development of biotic patterns across a landscape or seascape deepens, the likelihood that molecular genetic architecture within geographically isolated populations or closely related species sorts into discrete *reciprocally monophyletic* clades increases as well. Indeed, many phylogeographic studies are informative about events in biotic and Earth history several million years prior to the Pleistocene (Figure 7.1c), and often serve to refine the boundary between intraspecific and interspecific diversity through discovery of cryptic species or biogeographic events (ISI Web of Science search, 1993–2005, using "phylogeograph* and cryptic" revealed 148 studies). Of course, the pace of transformation of an ancestral *phylogroup* into two or more reciprocally monophyletic phylogroups is strongly associated with population genetic parameters that influence the rate of sorting of ancestral polymorphism and fixation of new mutations. This prevents the establishment of a universal metric across organisms for the timeframe within which we expect to discover reciprocally monophyletic phylogroups. Even so, one can identify the threads of an empirical, conceptual, and analytical bridge between phylogeography and the rest of historical biogeography at the nexus between population genetic and phylogenetic patterns of divergence among geographically structured populations.

FROM SINGLE-TAXON TO COMPARATIVE PHYLOGEOGRAPHY

Much as historical biogeographers did several decades ago, practitioners of phylogeography have increasingly recognized the need to examine congruence across multiple, co-distributed taxa to differentiate general events (e.g., vicariance or geo-dispersal; Lieberman, 2004) from taxon-specific events (e.g., long-distance dispersal and speciation of peripheral isolates; Bermingham & Moritz, 1998; Arbogast & Kenagy, 2001; Riddle & Hafner, 2004). *Comparative phylogeography* is defined, in part, as "the geographical comparison of evolutionary subdivision across multiple co-distributed species or species complexes" (Arbogast & Kenagy, 2001). The implication of phylogenetic subdivision produces conceptual affinities with area-based historical biogeography and, indeed, comparative phylogeographic studies often address vicariance vs. dispersal issues in a straightforward phylogenetic fashion. However, these studies also investigate population histories from a statistical phylogeographic approach when gene lineages in isolated populations have not yet

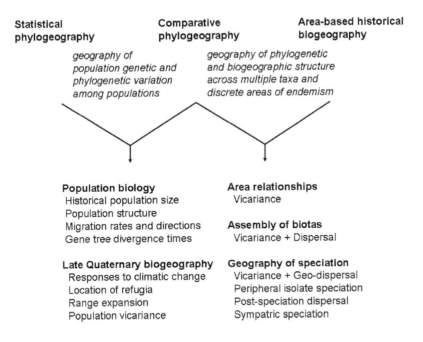

Statistical phylogeography	Comparative phylogeography	Area-based historical biogeography
geography of population genetic and phylogenetic variation among populations	*geography of phylogenetic and biogeographic structure across multiple taxa and discrete areas of endemism*	

Population biology
Historical population size
Population structure
Migration rates and directions
Gene tree divergence times

Late Quaternary biogeography
Responses to climatic change
Location of refugia
Range expansion
Population vicariance

Area relationships
Vicariance

Assembly of biotas
Vicariance + Dispersal

Geography of speciation
Vicariance + Geo-dispersal
Peripheral isolate speciation
Post-speciation dispersal
Sympatric speciation

FIGURE 7.2 A depiction of objectives and questions most appropriately addressed within the conceptually and analytically different arenas of statistical phylogeography, and area-based historical biogeography (see text for discussion), and illustrating the "bridging" position of comparative phylogeography.

sorted to a state of reciprocal monophyly (e.g., Rocha et al., 2002; Wares, 2002; Bremer et al., 2005; Lourie et al., 2005). Given the ability of comparative phylogeography to incorporate methods and approaches from both statistical phylogeography and area-based historical biogeography, it promises to form the basis for developing more robust linkages between these approaches (Figure 7.2) than currently exist.

As with area-based historical biogeography, comparative phylogeography offers a predictive framework in which an initial pattern of congruence across co-distributed taxa can be continually updated through addition of new co-distributed taxa and re-evaluation of patterns and explanations. An excellent example involves the postulated existence of a Pleistocene seaway across the Baja California Peninsula of Mexico. Flooding of the midpeninsula has been hypothesized since 1921, but was firs invoked to explain genetic discontinuities among peninsular species by Upton & Murphy (1997). Studying the side-blotched lizard (*Uta*), Upton & Murphy (1997) marshalled supportive evidence from the distribution and genetics of marine organisms in adjacent oceanic and gulf waters. They hypothesized a date of about 1 million years for the existence of this seaway separating the Peninsular North and Peninsular South areas (Figure 7.3). A comparative phylogeographic analysis of co-distributed mammalian, reptilian, and amphibian taxa provided evidence for a congruent break between southern and northern populations (Riddle et al., 2000) in the vicinity of

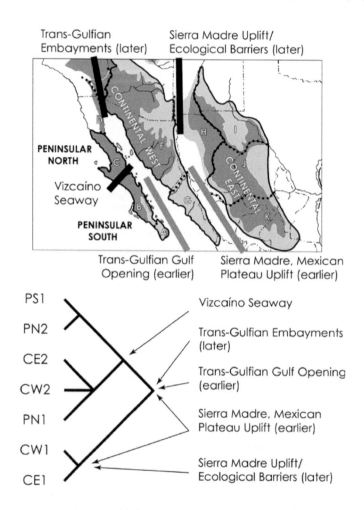

FIGURE 7.3 Summary diagram from a recent study using a step-wise approach to integrate phylogeographic and historical biogeographic analyses into a comprehensive analysis of biotic history in the warm deserts of western North America (summarized in text). The map shows areas of distribution (A–K), areas of endemism (Peninsular South, Peninsular North, Continental West, Continental East), and several postulated Neogene vicariant events. The area cladogram is a secondary BPA tree summarizing postulated historical events across these areas of endemism. (See Riddle & Hafner 2006a, b for details).

the postulated "Vizcaíno Seaway," which was consistent with evidence available at that time from additional reptilian (Upton & Murphy, 1997; Rodriguez-Robles & De Jesus-Escobar, 2000) and avian (Zink et al., 2001) taxa. Subsequently, various studies have confirmed the geographic generality of this southern vs. northern split in a wide variety of taxa, including more reptiles (Murphy & Aquirre-Leon, 2002; Lindell et al., 2005), spiders (Crews & Hedin, 2005), and marine fish (Bernardi et al., 2003; Riginos, 2005). Grismer (2002) has challenged invocation of a Pleistocene transpeninsular seaway, citing a lack of geological evidence (reviewed

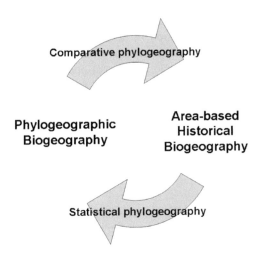

FIGURE 7.4 Illustration of one depiction of the synergistic and reciprocal nature of phylo-geographic biogeographic and area-based historical biogeographic perspectives and approaches, in which comparative and statistical phylogeography provide the hypothesis generating and hypothesis testing bridges, respectively, between the two approaches.

in Jacobs *et al.* 2004). Grismer (2002) further maintains that abrupt ecological changes in the midpeninsula are sufficient to explain genetic differentiation of terrestrial species. New evidence for marine fish (Riginos, 2005) provides particularly compelling support for a former seaway. Whereas an incomplete seaway, perhaps plugged with an active volcanic field or an abrupt ecological shift, would have limited dispersal of terrestrial forms up and down the peninsula, this would not account for disjunct marine distributions (Hafner & Riddle, 2005). The specifi timeframe in which such a seaway might have existed (e.g., Lindell et al., 2005; Riginos, 2005) and alternative explanations for north/south divergence across taxa (Grismer 2002; Crews & Hedin, 2006) are being actively explored. Nevertheless, this case study illustrates the value of a comparative phylogeographic approach in providing an appropriate framework for developing, recognizing, and testing hypotheses of historical divergence in a co-distributed biota. One weakness of the framework for addressing the generality of this pattern at this stage would be recognized clearly by area-based historical biogeographers: it reduces to only a two-area statement of area-relationships, which is meaningless within a cladistic biogeography framework that requires a minimum of three areas for postulating historical relationships — although panbiogeographers would consider it an acceptable general track.

In summary, we believe that a reconciliation of the concern over the validity of phylogeography within historical biogeography must recognize that questions explored under a phylogeographic rubric overlap strongly with those that form the core of the more time-honored, area-based historical biogeography, but also extend beyond the traditional envelope of that discipline. Yes, much of phylogeography is designed explicitly to investigate single-taxon dispersal hypotheses and utilizes sampling designs and statistical approaches appropriate for addressing questions such as population range expansion, locations of Late Quaternary refugia, and prevailing directions of migration between genetically divergent populations. However, within the multi-faceted phylogenetic and population genetic framework of phylogeography, the development and increasing application of comparative phylogeographic approaches provides a conceptual and empirical bridge between the statistical phylogeography that has arisen from a population genetic arena, and a more distinctly phylogenetic, area-based historical biogeography (Figure 7.2). Explicit attention to building such a synthetic perspective has only recently begun to enter the biogeographic literature.

TOWARD AN INTEGRATION OF PHYLOGEOGRAPHY AND HISTORICAL BIOGEOGRAPHY

Few studies have analyzed phylogeographic data within a cladistic biogeographic framework. Taberlet *et al.* (1998) used an earlier version of Brooks parsimony analysis (primary BPA) to postulate glacial-age refugia and routes of post-glacial colonization among 10 co-distributed taxa in Europe. Bermingham & Martin (1998) used COMPONENT 2.0 to address area-congruence across three genera of Neotropical fish. Hall & Harvey (2002) used a matrix representation with parsimony approach to generate a general area cladogram. The latter study combined a phylogeographic-scale cladogram from a neotropical butterfly genus with area cladograms derived from a wide range of published area cladograms of vertebrate taxa, most of which were not generated from phylogeographic data.

As has been recognized previously (Morrone & Crisci, 1995; Andersson, 1996; Crisci et al., 2003), questions that need to be addressed in a historical biogeographic analysis can be arranged sequentially, beginning with careful attention to the biotic and landscape scope of a study. We have suggested (Riddle & Hafner, 2006a, b) that phylogeographic and historical biogeographic analyses can be employed interactively across a series of sequential steps in an analysis of a biota in which divergence and distributional patterns likely occurred during the Neogene. This protocol and rationale, summarized briefly below, are discussed fully and illustrated by Riddle & Hafner (2006a and Figure 3 therein).

A wide range of studies over the past 25 years collectively have now provided convincing evidence that the appropriate units of analysis in a Neogene timeframe will, in most cases, be derived through phylogeographic-scale data and a phylogeographic sampling design. Phylogeographic-scale data include sequencing or otherwise diagnostic character state variation within rapidly evolving and rapidly sorting DNA partitions. Phylogeographic sampling involves multiple individuals from multiple sites distributed across the geographic range of a taxon. This sampling strategy is necessary to characterize accurately the geographic distribution of what are often morphologically cryptic phylogenetic lineages (i.e., phylogroups or evolutionarily significant units) within and among geographic distributional areas, and to provide an adequate database for statistical phylogeographic hypothesis testing. In our case study (Riddle & Hafner, 2006a, b), we initially identified 16 distributional areas (10 of which are shown as areas A through K in Figure 7.3) for taxa in 22 clades of mammals, birds, reptiles, amphibians, and plants with decidedly "warm desert" ecological affinities, distributed primarily across the major warm deserts of southwestern North America (Figure 7.3).

Distributional areas are not, by themselves, robustly recognizable as areas of endemism. Additional analysis of the degree of distributional overlap among taxa, as well as recognition of barriers and filte -barriers among areas and phylogenetic and distributional congruence among taxa, are also required (Lomolino et al., 2006). We employed an iterative parsimony analysis of endemicity (PAE) proce- dure to assess distributional congruence of phylogroups from the 22 cladograms across distributional areas. Based on the scale of resolution available in the phy- logeographic data, this resulted in a set of four core warm desert areas of endemism (Figure 7.3).

Once areas of endemism were identified, we selected primary and secondary Brooks parsimony analysis (BPA; Brooks et al., 2001) as an appropriate method for analyzing area relationships across the four core areas. Our experience with the North American warm-desert biota led us to suspect a high level of idiosyncratic responses of taxa to postulated vicariant events and variable rates of historical dispersal across areas, which together are likely to result in high levels of reticulation across areas. Primary and secondary BPA were designed to disentangle reticulation across areas by resolving area cladograms into a set of area relationships derived through postulated vicariance, sympatric speciation, post-speciation dispersal, spe- ciation of peripheral isolates, and extinction events. Our secondary BPA tree (Figure 7.3) is consistent with the Neogene geological history of southwestern North America and provides a realistic depiction of the temporal continuum inherent in the tectonically driven transformation of a landscape (Riddle and Hafner, 2006a, b). Moreover, it provides an objective depiction of major sources of reticulation in the area-relationships depicted on the primary BPA cladogram: the equivalent role of the Continental West area in diversification and dispersal to the west (peninsular areas) and east (Continental East area); and a similarly central role for the Peninsular North area between Peninsular South and Continental West. Recently, Wojcicki & Brooks (2004, 2005) have introduced a new algorithm (PACT = phylogenetic analysis for comparing trees) that builds upon BPA and purportedly "embodies all of the strong points and none of the weaknesses of previously proposed methods of historical biogeography." We eagerly await the availability of this algorithm in a computer program package so that its effica y can be evaluated against other available approaches. We believe that perhaps the most powerful feature of our fi e-step approach (Riddle & Hafner 2006a, b) is the heuristic nature of the resulting area cladogram, which provides a number of interesting hypotheses that can then be addressed by turning to a statistical phy- logeographic arena. For example, events that might otherwise be interpreted by historical biogeographers based on cladogenetic structure alone as "sympatric speciation" actually comprise two alternative hypotheses: sympatric speciation or embedded allopatric speciation, possibly with subsequent range expansion of one or both sister taxa. These alternatives can be addressed through a variety of available statistical phylogeographic analyses.

One pattern of great interest in our study involves the population history of taxa on the Baja California Peninsula. While nine phylogroups are currently endemic to the Peninsular South area, another nine occur across both of the peninsular areas. Thus, because of the high level of endemicity in the southern area and a postulated vicariant event that putatively is associated, at least in part, with this pattern (Figure 7.3), we consider either post-barrier range expansion (primarily from south to north) or ancestrally widespread distributions as two alternative hypotheses to explain the currently widespread distribution of the nine taxa across southern and northern areas. Again, these hypotheses are highly tractable by using statistical phylogeographic analyses to assess the likelihood of each alternative. In fact, previous phylogeographic studies of two of these taxa have provided evidence in favor of recent northward range expansion from the Peninsular South area (Zink et al., 2000; Nason et al., 2002). While those studies were conducted within a typical single-taxon phylogeographic framework, the synthetic approach we have presented here illustrates the predictive power of working interactively between phylogeography and area-based historical biogeography. By recognizing the strong signal of endemism and separating the histories of the southern and northern peninsular areas, we have constructed an objective framework for entertaining the hypothesis that post-barrier range expansion has not only been a general feature in the Neogene history of this biota, but is primarily asymmetrical in direction.

FUTURE DIRECTIONS

We foresee a rapid strengthening of the developing bridge between statistical phylogeography and area-based historical biogeography. The explosive growth of single-taxon phylogeographic studies should inevitably lead to additional comparative phylogeographic studies of regions such as our analyses of North American warm deserts (Riddle & Hafner 2006a, b). These, in turn, will allow comparative studies among diverse regions that address questions of deeper temporal scale and broader geographic scale (e.g., Lessa et al., 2003). We anticipate emergence of new tools to optimize the synergism and reciprocal strengths (Figure 7.3) between, on the one hand, population-genetic approaches that examine detailed dynamics of individual taxa in local environments and, on the other hand, area-based historical biogeographic approaches that seek to summarize regional biotic histories.

ACKNOWLEDGEMENTS

We thank the National Science Foundation for support of research reported here through grants DEB-9629787 and 0237166 (to B.R. Riddle) and DEB-9629840 and 0236957 (to D.J. Hafner).

REFERENCES

Andersson, L. (1996). An ontological dilemma: epistemology and methodology of historical biogeography. *Journal of Biogeography,* 23, 269–277.

Arbogast, B.S. & Kenagy, G.J. (2001). Comparative phylogeography as an integrative approach to historical biogeography. *Journal of Biogeography,* 28, 819–825.

Avise, J.C., Arnold, J., Ball, R.M., Bermingham, E., Lamb, T., Neigel, J.E., Reeb, C.A. & Saunders, N.C. (1987). Intraspecific phylogeography — the mitochondrial-DNA bridge between population-genetics and systematics. *Annual Review of Ecology and Systematics,* 18, 489–522.

Avise, J. (2000). *Phylogeography: The History and Formation of Species.* Harvard University Press, Cambridge, MA.

Bermingham, E. & Martin, A.P. (1998). Comparative mtDNA phylogeography of neotropical freshwater fishes: testing shared history to infer the evolutionary landscape of lower Central America. *Molecular Ecology,* 7, 499–517.

Bermingham, E. & Moritz, C. (1998). Comparative phylogeography: concepts and applications, *Molecular Ecology,* 7, 367–369.

Bernardi, G., Findley, L. & Rocha-Olivares, A. (2003). Vicariance and dispersal across Baja California in disjunct marine fish populations, *Evolution,* 57, 1599–1609.

Bremer, J.R.A., Vinas, J., Mejuto, J., Ely, B. & Pla, C. (2005). Comparative phylogeography of Atlantic bluefin tuna and swordfish: the combined effects of vicariance, secondary contact, introgression, and population expansion on the regional phylogenies of two highly migratory pelagic fishes. *Molecular Phylogenetics and Evolution,* 36, 169–187.

Brooks, D.R. (2004). Reticulations in historical biogeography: the triumph of time over space in evolution. In *Frontiers in Biogeography: New Directions in the Geography of Nature* (ed. by Lomolino, M.V. & Heaney, L.R.), pp. 125–144. Sinauer Associates Inc., Sunderland, MA.

Brooks, D.R., van Veller, M.G.P. & McLennan, D.A. (2001). How to do BPA, really. *Journal of Biogeography,* 28, 345–358.

Brooks, D.R., Dowling, A.P.G., van Veller, M.G.P. & Hoberg, E.P. (2004). Ending a decade of deception: a valiant failure, a not-so-valiant failure and a success story. *Cladistics — The International Journal of the Willi Hennig Society,* 20, 32–46.

Crews, S.C. & Hedin, M. (2006). Studies of morphological and molecular phylogenetic divergence in spiders (Araneae: *Homalonychus*) from the American southwest, including divergence along the Baja California Peninsula. *Molecular Phylogenetics and Evolution,* 38, 470–487.

Crisci, J.V., Katinas, L. & Posadas, P. (2003). *Historical Biogeography: an Introduction.* Harvard University Press, Cambridge, MA.

Donoghue, M.J. & Moore, B.R. (2003). Toward an integrative historical biogeography. *Integrative and Comparative Biology,* 43, 261–270.

Ebach, M.C. (2001). Extrapolating cladistic biogeography: A brief comment on van Veller *et al.* (1999, 2000, 2001). *Cladistics — The International Journal of the Willi Hennig Society,* 17, 383–388.

Ebach, M.C. & Humphries, C.J. (2002). Cladistic biogeography and the art of discovery. *Journal of Biogeography,* 29, 427–444.

Ebach, M.C. & Humphries, C.J. (2003). Ontology of biogeography. *Journal of Biogeography,* 30, 959–962.

Grismer, L.L. (2002). A re-evaluation of the evidence for a mid-Pleistocene mid-peninsular seaway in Baja California: a reply to Riddle *et al. Herpetological Review,* 33, 15–16.

Hafner, D.J. & Riddle, B.R. (2005). Mammalian phylogeography and evolutionary history of northern Mexico's deserts. *Biodiversity, Ecosystems, and Conservation in Northern Mexico* (ed. by Catron, J.-L.E., Ceballos, G. & Felger, R.S.), pp. 225–245. Oxford University Press, New York.

Hall, J.P.W. & Harvey, D.J. (2002). The phylogeography of Amazonia revisited: new evidence from riodinid butterflies. *Evolution*, 56, 1489–1497.

Hewitt, G.M. (2004). Genetic consequences of climatic oscillations in the Quaternary. *Philosophical Transactions of the Royal Society of London Series B — Biological Sciences*, 359, 183–195.

Humphries, C.J. (2000). Form, space and time; which comes first? *Journal of Biogeography*, 27, 11–15.

Humphries, C. & Ebach, M.C. (2004). Biogeography on a dynamic earth. In *Frontiers of Biogeography: New Directions in the Geography of Nature* (ed. by Lomolino, M.V. & Heaney, L.R.), pp. 67–86. Sinauer Associates Inc., Sunderland, MA.

Jacobs, D.K, Haney, T.A. & Louie, K.D. (2004). Genes, diversity, and geologic process on the Pacific Coast. *Annual Review of Earth and Planetary Sciences*, 32, 601–652.

Knowles, L.L. (2004). The burgeoning field of statistical phylogeography. *Journal of Evolutionary Biology*, 17, 1–10.

Lessa, E.P, Cook, J.A. & Patton, J.L. (2003). Genetic footprints of demographic expansion in North America, but not Amazonia, following the Late Pleistocene. *Proceedings of the National Academy of Sciences USA*, 100, 10331–10334.

Lieberman, B.S. (2004). Range expansion, extinction, and biogeographic congruence: a deep time perspective. In *Frontiers of Biogeography: New Directions in the Geography of Nature* (ed. by Lomolino, M.V. & Heaney, L.R.), pp. 111–124. Sinauer Associates, Inc., Sunderland, MA.

Lindell, J., Mendez-de la Cruz, F.R. & Murphy, R.W. (2005). Deep genealogical history without population differentiation: discordance between mtDNA and allozyme divergence in the zebra-tailed lizard (*Callisaurus draconoides*). *Molecular Phylogenetics and Evolution*, 36, 682–694.

Lomolino, M.V., Riddle, B.R. & Brown, J.H. (2006). *Biogeography*, 3rd ed. Sinauer Associates, Inc., Sunderland, MA.

Lourie, S.A., Green, D.M. & Vincent, A.C.J. (2005). Dispersal, habitat differences, and comparative phylogeography of Southeast Asian seahorses (Syngnathidae: *Hippocampus*). *Molecular Ecology*, 14, 1073–1094.

Mateos, M. (2005). Comparative phylogeography of livebearing fishes in the genera *Poeciliopsis* and *Poecilia* (Poeciliidae: Cyprinodontiformes) in central Mexico. *Journal of Biogeography*, 32, 775–780.

Morrone, J.J. & Crisci, J.V. (1995). Historical biogeography — introduction to methods. *Annual Review of Ecology and Systematics*, 26, 373–401.

Murphy, R.W. & Aguirre-Léon, G. (2002). Nonavian reptiles; origins and evolution. In *A New Island Biogeography of the Sea of Cortés* (ed. by Case, T.J., Cody, M.L. & Ezcurra, E.), pp. 181–220. Oxford University Press, New York.

Nason, J.D., Hamrick, J.L. & Fleming, T.H. (2002). Historical vicariance and postglacial colonization effects on the evolution of genetic structure in *Lophocereus*, a Sonoran Desert columnar cactus. *Evolution*, 56, 2214–2226.

Nelson, G. (1974). Historical biogeography — alternative formalization. *Systematic Zoology*, 23, 555–558.

Riddle, B.R. (2005). Is biogeography emerging from its identity crisis? *Journal of Biogeography*, 32, 185–186.

Riddle, B.R. & Hafner, D.J. (2004). The past and future roles of phylogeography in historical biogeography. *Frontiers of Biogeography* (ed. by Lomolino, M.V. & Heaney, L.R.), pp. 93–110. Sinauer Associates, Inc., Sunderland, MA.

Riddle, B.R. & Hafner, D.J. (2006a). A step-wise approach to integrating phylogeographic and phylogenetic biogeographic perspectives on the history of a core North American warm deserts biota. *Journal of Arid Environments*, 65, 435–461.

Riddle, B.R. & Hafner, D.J. (2006b). Biogeografía y biodiversidad de los desiertos cálidos del norte de México y soroeste de Estados Unidos. *Genética y mamíferos mexicanos: presente y futuro* (ed. by Vázquez, E. & Hafner, D.J.), pp. 57–65. New Mexico Museum of Natural History and Science Bulletin, 32.

Riddle, B.R., Hafner, D.J., Alexander, L.F. & Jaeger, J.R. (2000). Cryptic vicariance in the historical assembly of a Baja California Peninsular Desert biota. *Proceedings of the National Academy of Sciences USA*, 97, 14438–14443.

Riginos, C. (2005). Cryptic vicariance in Gulf of California fishes parallels vicariant patterns found in Baja California mammals and reptiles. *Evolution*, 59, 2678–2690.

Rocha, L.A., Bass, A.L., Robertson, D.R. & Bowen, B.W. (2002). Adult habitat preferences, larval dispersal, and the comparative phylogeography of three Atlantic surgeonfishe (Teleostei: Acanthuridae). *Molecular Ecology*, 11, 243–252.

Rodriguez-Robles, J.A. & De Jesus-Escobar, J.M. (2000). Molecular systematics of new world gopher, bull, and pinesnakes (Pituophis: Colubridae), a transcontinental species complex. *Molecular Phylogenetics and Evolution*, 14, 35–50.

Siddall, M.E. (2005). Bracing for another decade of deception: the promise of Secondary Brooks Parsimony Analysis. *Cladistics*, 21, 90–99.

Siddall, M.E. & Perkins, S.L. (2003). Brooks parsimony analysis: a valiant failure. *Cladistics — The International Journal of the Willi Hennig Society*, 19, 554–564.

Taberlet, P., Fumagalli, L., Wust-Saucy, A.G. & Cosson, J.F. (1998). Comparative phylogeography and postglacial colonization routes in Europe. *Molecular Ecology*, 7, 453–464.

Templeton, A.R. (2004). Statistical phylogeography: methods of evaluating and minimizing inference errors. *Molecular Ecology*, 13, 789–809.

Templeton, A.R., Routman, E. & Phillips, C.A. (1995). Separating population structure from population history: a cladistic analysis of the geographical distribution of mitochondrial DNA haplotypes in the tiger salamander, *Ambystoma tigrinum*. *Genetics*, 140, 767–782.

Upton, D.E. & Murphy, R.W. (1997). Phylogeny of the side-blotched lizards (Phrynosomatidae: *Uta*) based on mtDNA sequences: Support for a midpeninsular seaway in Baja California. *Molecular Phylogenetics and Evolution*, 8, 104–113.

Van Veller, M.G.P. & Brooks, D.R. (2001). When simplicity is not parsimonious: *a priori* and *a posteriori* methods in historical biogeography. *Journal of Biogeography*, 28, 1–11.

Van Veller, M.G.P., Brooks, D.R. & Zandee, M. (2003). Cladistic and phylogenetic biogeography: the art and the science of discovery. *Journal of Biogeography*, 30, 319–329.

Wares, J.P. (2002). Community genetics in the Northwestern Atlantic intertidal. *Molecular Ecology*, 11, 1131–1144.

Wiens, J.J. & Donoghue, M.J. (2004). Historical biogeography, ecology and species richness. *Trends in Ecology & Evolution*, 19, 639–644.

Wojcicki, M., & Brooks, D.R. (2004). Escaping the matrix: a new algorithm for phylogenetic comparative studies of co-evolution. *Cladistics*, 20, 341–361.

Wojcicki, M. & Brooks, D.R. (2005). PACT: an efficient and powerful algorithm for generating area cladograms. *Journal of Biogeography*, 32, 755–774.

Zheng, X.G., Arbogast, B.S. & Kenagy, G.J. (2003). Historical demography and genetic structure of sister species: deermice (*Peromyscus*) in the North American temperate rain forest. *Molecular Ecology,* 12, 711–724.

Zink, R.M., Barrowclough, G.F., Atwood, J.L. & Blackwell-Rago, R.C. (2000). Genetics, taxonomy, and conservation of the threatened California Gnatcatcher. *Conservation Biology,* 14, 1394–1405.

Zink, R.M., Kessen, A.E., Line, T.V. & Blackwell-Rago, R.C. (2001). Comparative phylogeography of some aridland bird species. *Condor,* 103, 1–10.

8 Are Plate Tectonic Explanations for Trans-Pacific Disjunctions Plausible? Empirical Tests of Radical Dispersalist Theories

Dennis McCarthy

ABSTRACT

As has been known for decades, *hundreds* of identical or closely-related terrestrial tetrapods, freshwater fish, shallow-water marine fauna, and other taxa of restricted vagility, both fossil and extant, are separated by the vast Pacific Since 1988, a number of researchers have compared vicariance vs. dispersal explanations for these trans-Pacific disjunctions and have concluded that vicariance should be preferred. This is controversial because it conflicts with standard plate tectonic models of the history of the Pacific Those who oppose these determinations of vicariance have instead assumed repeated incidences of trans-Pacific, sweepstakes-crossings of poor-dispersing taxa and cross-Pangaean range expansions that are obscured by wide-scale patterns of fossil absences. These dispersalist hypotheses conflict with the following biogeographical facts and analyses:

The Pitcairn Proof: Kingston *et al.*'s (2003) detailed investigation of range data and the dispersal method of each of the 114 species of flora from the Pitcairn group corroborates the view that taxa that can manage extremely long-distance, trans-marine colonization are very likely to be wide ranging.

The du Toit Principle: The currently uncontroversial explanation of pre-Jurassic, trans-Atlantic and trans-Indian disjunctions, first put forth by revolutionaries like Alexander du Toit and Alfred Wegener, imply that trans-oceanic disjunctions of such taxa are most reasonably explained by vicariance — not by dispersal and fossil absence.

The False-Positive Challenge: Defenders of the dispersalist hypotheses have not provided a single uncontroversial example of coordinated, systematic, long-distance dispersals (or massive range expansions and contractions) among a wide variety of taxa leading to a false-positive signal for vicariance. Vicariant explanations may be firmly rejected for countless pairs of regions, say, Wisconsin and Sri Lanka, yet are there any such widely separated pairs that *exclusively* share a wide variety of taxa?

The combination of analyses presented here lend additional support to those otherwise uncontroversial biogeographical principles that conflict with plate tectonics, confirming the conclusions of numerous researchers who have studied the trans-Pacific disjunctions over the last few decades — the ocean crossing scenarios and patterns of missing fossils, first put forth to rescue continental stabilism and recycled today to save plate tectonics, do not just *seem* implausible, they are implausible.

INTRODUCTION / THE DU TOIT DENOUEMENT

The notion of random, and sometimes two-way, 'rafting' across the wide oceans ... evinces ... a weakening of the scientific outlook, if not a confession of doubt from the viewpoint of organic evolution.

To argue that such southern disjunctive distribution is due to colonisation from the north through forms not yet discovered in the Holarctic region, is neither scientific nor fair ... .

The preceding quotes adorn Alexander du Toit's classic riposte (Du Toit, 1944) to G.G. Simpson, who had put forth a now-infamous defense of the then-mainstream geological view of continental stabilism (Simpson, 1940, 1943). Simpson, like many biogeographers and geologists of his time, ridiculed the notion of closed oceans and attempted to explain the trans-Atlantic and trans-Indian disjunctions of poorly dispersing fossil taxa through a combination of cross-ocean raftings, island hopping, and convenient patterns of fossil absences. Although nearly all scientists now agree that du Toit was correct and that Simpson's arguments were incredible, we still fin ourselves, 60 years later, in the midst of this same debate. This time the disjunctions involve poor-dispersing taxa of a much greater quantity and variety — and a marine barrier significantly more imposing.

As has been known for decades, *hundreds* of identical or closely-related, terrestrial tetrapods, freshwater fish, shallow marine fauna, and other taxa of restricted vagility, both fossil and extant, are separated by the vast Pacific (Croizat, 1958; Hennig, 1960; Brundin, 1966; Shields, 1979, 1998; Ager, 1986; Newton, 1988; Matile, 1990; Grehan, 1991; Craw et al., 1999; Glasby, 1999; Heads, 1999; Humphries & Parenti, 1999; Sequeira & Farrell, 2001; Cranston, 2003; McCarthy, 2003). Since 1988, a number of researchers have compared vicariance vs. dispersal explanations for these trans-Pacific disjunctions and, like du Toit, have concluded that vicariance should be preferred (e.g., Ager, 1986; Matile, 1990; Shields, 1998; Glasby, 1999; Heads, 1999; Cranston, 2003; McCarthy, 2003). The rationale behind their conclusions is not controversial and is referred to here as "du Toit's Principle".

Du Toit's Principle: Any geological theory that assumes the continuous existence of a hypothetical ocean between regions that exclusively share a wide variety of poor-dispersing taxa necessitates a myriad of additional assumptions involving improbable cross-ocean dispersals or circuitous range expansion with convenient fossil absences. As the multiplying of such unlikely hypotheses should be avoided in scientific theories, another geological theory that places the taxa in proximity provides a more parsimonious explanation for these distributions and should be preferred.

Examples clarify. The basal, sister taxon relationship of the Californian iguanid *Dipsosaurus* with the Fijian *Brachylophus* (Sites et al., 1996) and the "spectacularly large" (at least 1.5 m in length) Fijan fossil iguanid *Lapitiguana impensa* (Pregill & Worthy, 2003) necessitates, according to plate tectonics (PT), a rafting trip of a gravid female or Adam and Eve pair or a clutch of eggs across a minimum of 8000 km of open ocean from the Western Americas to Fiji (Cogger, 1974; Gibbons, 1981). This trip is nearly three times longer than the now-forsaken, trans-Atlantic rafting trips of larger vertebrates that Simpson supposed in his defense of continental stabilism. And it is simply one of hundreds of additional hypotheses required by PT in order to explain trans-Pacific disjunctions, each one calling for a new raft (e.g., O'Foighil et al., 1999) or an additional uninterrupted chain of islands or seamounts to hop (e.g., Newton, 1988; Rieppel et al., 2000) or another cross-Pangean distribution with careful avoidance of fossil sites (e.g., Briggs, 2004). In PT, the biogeographical problems solved by closing the Atlantic and Indian still confound in the Pacific

The vicariance interpretation of expanding Earth (EE), in contrast, explains all the disjunctions simply and with a single cause: the opening and expansion of the Pacific All ocean basins exclusively comprise crust younger than 200 mya (Figure 8.1a–d), and the continental margins that bracket each basin, Pacifi included, fit together like puzzle pieces (Scalera, 1993; McCarthy, 2003). This evidence is consistent with the EE view that all ocean basins were closed, or nearly so, in pre-Jurassic time, and all large, deep marine environments consisted of epicontinental seas. According to EE, in the Late Cretaceous and Early Tertiary, the South Pacific was significantly narrower, not wider, permitting exchange of taxa of more limited mobility (Figure 8.1c). Plate tectonics, in contrast, supposes the existence and then complete subduction of a pre-Pacific superocean, Panthalassa (Figure 8.2). According to plate tectonics, the pre-Pacific Basin was considerably larger in the Mesozoic and has steadily become narrower.

Each of the referenced researchers (e.g., Shields, 1979, 1998; Ager, 1986; Matile, 1990; Glasby, 1999; Heads, 1999; Cranston, 2003; McCarthy, 2003, 2005a) examined both sides of this vicariance-dispersal debate and reasoned that a theory that brought all the disjunct taxa in proximity provided a more reasonable explanation for these distributions than plate tectonics, which needed to incorporate so many new and fantastic hypotheses. Biogeographically speaking, these analyses are not controversial.

Since the publication of these papers, Briggs (2004), whose contributions to the field of biogeography are legion, became the first researcher to provide a reply that appears to challenge this argument. Briggs' knowledge and expertise in biogeography are beyond dispute, and many of his keen rebuttals to McCarthy (2003) have also been put forth by geologists, also of inspiring credentials. It is hoped that the detailed nature of this response gives the appearance of thoroughness, not confrontation. Plate tectonics has guided the pen of paleocartographers for more than three decades, and it is important to clarify the strength of the evidence against these views.

Ages of Ocean Basin Crust

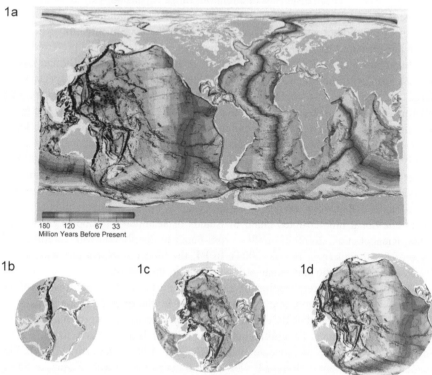

180   120   67 33
Million Years Before Present

Late Triassic   Late Cretaceous   Present

FIGURE 8.1 (a) From an oceanic crustal age poster from the National Oceanic and Atmospheric Administration (NOAA) National Geophysical Data Center (based on Mueller et al., 1993). All oceanic crust is < 200 Mya old, and most of the South Pacific formed <40 Mya. (b–d) are EE globes (McCarthy 2005a) of the Pacific hemisphere that follow from the crustal age data. Some distortion should be expected due to transference of the flat map projections of (a) to a global view. This simple and constrained development of the Pacific plate is predictive both of the timing and location of trans-Pacific disjunctions. (b) All the disjunct trans-Pacific fossil taxa from the Late Triassic-Early Jurassic are reunited (Figure 8.2). (c) The Late Cretaceous and Early Tertiary narrow marine gap between New Zealand and southern South America — as well as between central/tropical South America and Melanesian regions like New Guinea, New Caledonia, and Fiji (a biotic province termed "Melanamerica" by Malte Ebach, pers. comm.). This would allow interchange of taxa capable of *narrow* transmarine dispersal — but not oceanic jumps (e.g., New Zealand's frog *Leiopelma* and lizardlike tuatara, New Caledonia's nearly flightless kagu, Fiji's iguanas).

As stated, Briggs *appeared* to dispute the arguments of the vicariance group, but is he actually contending that it is more parsimonious to explain the disjunctions of poor-dispersing taxa by hypothesizing a superocean between them? Is he arguing, for example, that it is more likely for an iguanid to be able to raft >8000 km than negotiate a narrow seaway? Of course not. In his book *Biogeography and Plate*

Conventional Paleomap of Upper Triassic

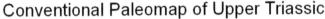

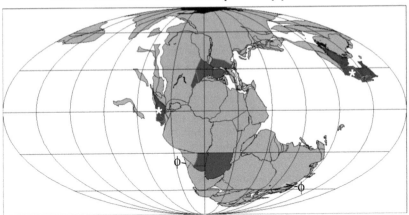

Shaded regions denote trans-oceanic disjunctions of poor-dispersers

*	**W. N. America – East Asia**		**Europe – N. America**
φ	**Australia – S. America**		**S. Africa – S. America**

FIGURE 8.2 Mollweide plate tectonic paleomaps from the Early Jurassic (from Schettino & Scotese, 2001). The shaded continental areas represent a few of the trans-oceanic regions that exclusively shaded an inordinate number of Early Jurassic and/or pre-Jurassic taxa (see text). As no current ocean basin crust is >200 mya old, all seafloor depicted is hypothetical. Index of similarity analyses for tetrapod families for Late Triassic-Early Jurassic show North America–East Asia (64) have the same index as South America–South Africa (64) and a higher index than North America-Europe (55) (Hotton, 1992). The polar view shows the extreme distance of the top of the world route that Briggs (2004) advocates for communication of the terrestrial trans-Panthalassa taxa. Many of these regions, like Russia, Europe, and eastern North America, are fossiliferous, but the relevant taxa do not appear in these regions (Hotton, 1992; Shields, 1998). Moreover, lack of cross-Pangea marine corridors in the Late Triassic still necessitates trans-Panthalassa island hopping for shallow marine reptiles, like pistosaurs and thalattosaurs (Rieppel et al., 2000), and seamount hopping for all other shallow water marine fauna (Newton, 1988).

Tectonics, Briggs (1987) clearly advances du Toit's principle repeatedly, writing, for example: "There is no doubt that, in the Triassic, South America was closely joined to the other [Gondwanan] continents. In fact, at this time its vertebrate fauna was closer to that of Africa and Europe rather than North America" (Briggs, 1987, p. 133). In the same work, Briggs challenges a hypothesis of a large Tethys barrier separating India from Africa during the Cretaceous for similar reasons: "There is no way in which India could exhibit such a close biological relationship to Africa and at the same time have been isolated from it for tens of millions of years" (Briggs,

1987, p.137). As recently as 2003, Briggs followed up on this theme and rejected long-standing PT depictions of India in the Late Cretaceous, because they would require iguanid lizards and other non-volant vertebrates to cross a wide marine barrier from Eurasia. Briggs (2003) instead argues that India was significantly larger at that time, large enough to be close to Madagascar, Africa, and Asia simultaneously, permitting commerce among vertebrates. In brief, he discounts the possibility of island hopping schemes or rafting events and narrows the Tethys gap separating Eurasia, India, and Madagascar. (Interestingly, a narrow Tethys is predicted naturally in expanding Earth theory — and various EE theorists have been putting forth arguments for a narrow Tethys, biogeographical and otherwise, for many years [Carey, 1988; Shields, 1996]). Quoting Briggs (2003):

> As noted, a variety of northern forms had begun to appear in the late Cretaceous–early Paleocene strata of India. Concurrently, the southern tip of India must have remained close to Madagascar as indicated by the similarity of the vertebrate faunas of the terminal Cretaceous. Also, if the iguanid lizards and boid snakes of Madagascar had migrated from Asia they would have done so at about this time and there were, in India, dinosaurs related to those of Madagascar. None of these relationships could have been achieved, had India still been located in the middle of the Indian Ocean (Briggs, 2003, p. 386).

While followers of EE must applaud Briggs for joining the fraternity of research-ers who have challenged geological convention and closed a hypothetical oceanic gap due to exorbitant disjunctions, we still must highlight a larger point: if it is unreasonable to expect non-volant vertebrates like iguanids to reach the middle of the Tethys ocean, it is even more so to contend that they navigated the entire width of the Panthalassa/Pacific

Briggs certainly cannot and does not disagree with du Toit's principle, that is, the rationale of the cited trans-Pacific researchers. Evidently, it is not the *analyses* that he finds objectionable; it is the *conclusion*. Briggs rejects EE for reasons non-biogeographical. (Those points will be addressed later). Much like Simpson, another indisputable expert in biogeography, he does not, in this particular instance, use distributional facts to test conventional geological views. He subordinates distribu-tional facts to conventional geological views. Briggs uses standard PT paleomaps to govern his biogeographical explanations, suggesting as many hypotheses as nec-essary to reconcile fact with theory. As noted, the explanations required by plate tectonics are precisely the same as those offered by Simpson in support of a pre-Jurassic Atlantic, but given that they involve Panthalassa and Pangea, they are even more extreme.

SHOULD OCEAN-CROSSING TAXA
BE WIDE-RANGING?

Briggs' 2004 response sidestepped essentially all of the tropical and many of the southern disjunctions of extant taxa, including the Odyssean voyages required by *Brachylophus* and *Ostrea chilensis*, because they "cannot be used as evidence for the alignment of Triassic continents or islands" (Briggs, 2004, p. 856). While this is correct, no one was making that argument. In all cases, the evidence is consistent with the possibility that these disjunctions occurred during or prior to the Early Tertiary when, according to EE, the tropical and southern Pacifi was significantly narrower. As the crustal age data reveal (Figure 8.1a–d), the Pacific "un-zippered" open from north to south, maintaining direct terrestrial connections among South America, Australia, and New Zealand into the Cretaceous, and remaining a narrow seaway in the south until as late as the Eocene (McCarthy, 2003). The matching trans-Pacific geological outlines and taxa suggest a "trans-Pacific zipper effect", and the regions that were previously locked together are still apparent.

Dispersalists attempting to harmonize trans-Pacific disjunctions with geological tradition must confront two incontrovertible facts:

1. Oceanic and continental islands provide insurmountable empirical evidence as to the types of taxa that can cross wide marine barriers (>3000 km), those that can struggle past narrow marine barriers (<700 km), and those that cannot seem to cross any sort of marine barrier at all. For example, narrow marine breaks are prohibitive to large non-volant terrestrial vertebrates (e.g., Diamond, 1990), and except perhaps for geckos and skinks, wider gaps baffle the small ones.
2. Taxa that have traveled great distances to more remote oceanic regions tend to be wide ranging, usually occurring on other oceanic island groups.

Certainly, most biogeographers agree that long-distance dispersal adequately explains the distributions of many taxa on New Zealand, New Caledonia, and Fiji, and that similar trans-marine jumps are the predominant, if not exclusive, mechanism of colonization of the juvenile oceanic islands. Recently, McDowall (2004) has successfully argued this case by highlighting the examples of notable dispersals required by many Hawaiian plants and animals, a point which few deny. But just as Hawaii proves that certain taxa are capable of great voyages over marine barriers, it also confirms the vicariant biogeographer's point that many types of taxa are not. Hawaii has no native, non-volant, terrestrial vertebrates. The Hawaiian Islands also lack conifers, most bird families, and 14 of 30 recognized orders of insects. Before human introduction, Hawaii had maintained independence from the kingdom of the mosquitoes, and even the ubiquitous ant failed to annex it (e.g., Howarth et al., 1995; Loope, 1998; Jacobi, 2001). And Hawaii is merely half way across the Pacific, just 3900 km from the nearest continent. The dispersalist scenarios advocated by followers of PT require repeated negotiations of a marine barrier more than twice that, involving the types of taxa whose inability to cross wide marine barriers is implied by their conspicuous biophysical limitations and confirmed by their absence from remote islands like Hawaii.

Another implausibility confronting plate tectonics is that many of the taxa alleged to have accomplished these oceanic leaps occur in only a narrow range of locations, absent from all other proximal island groups and often confined to just two small regions that bracket the Pacific The problem this presents is elementary — if a taxon can conquer an >8000-km marine barrier, it should be able to reach locations significantly nearer. McDowall (2004) seemed to challenge this view, but this is likely the result of miscommunication, specifically a difference in use of certain terms. For example, "trans-oceanic" in McCarthy (2003) refers to a trip from one side of the Pacific Ocean to the other, but McDowall (2004) used the phrase "trans-oceanic" as a synonym for "trans-marine", as is quite common. McDowall (2004) writes: "Although formerly present only in Australia and islands to the north, even a diadromous freshwater fish has recently begun arriving in New Zealand [reference therein]. There is absolutely no reason why the relationships '[should be chaotic and wide ranging]', as McCarthy (2003) argued" (McDowall, 2004, p. 347). But McDowall's taxon is not inconsistent with McCarthy's point. Its sea-crossing (as opposed to *ocean*-crossing) capability is not only concordant with its diadromy, differentiating it from the vast majority of the significantly less vagile taxa cited by McCarthy, but is evidenced by its colonization of islands to the north. (The only diadromous fish referenced by McCarthy were galaxiids, and even they include the trans-Pacific disjunct sisters, *Brachygalaxias* and *Galaxiella,* which are freshwater limited [Waters et al., 2000]). McDowall is certainly right that this relatively narrow transmarine dispersal should not imply a distribution more wide-ranging than that. But this is irrelevant to the cross-Pacific jumps required by PT and the fact that taxa capable of such extraordinary trips should, in general, be capable of lesser feats.

The very premises of the dispersalist creed bolster this argument. Anemochory, zoochory, and hydrochory are a numbers game. Over the course of thousands of years, untold numbers of spores, seeds, larvae, fruit-eating birds, and small vertebrates clinging to flotsam will travel over or with the ocean, many eventually producing a successful colonization. Islands begin to green and buzz and chirp. Over the course of millions of years, taxa amenable to such processes over a certain distance will begin to occupy many of the suitable habitats within that distance. Even if Aeolos and Poseidon carry the floaters and drifters in a particular direction, the taxa will at least have a chance to occupy regions in that direction. This invalidates the only rationalization provided for the trans-oceanic jaunts of poor-dispersers, to wit, given enough time, even the extremely improbable can happen. The hypothesis that these disjunct taxa have crossed the entire width of the Pacific but have yet to colonize islands and regions considerably closer is improbable to begin with and *becomes more improbable with time.*

Empirical evidence is not lacking: the Pitcairn Island group lies roughly halfway between New Zealand and South America, ~4800 km from each coast, and its relatively recent geological formation (~16 mya for the oldest Island, Henderson, and perhaps <1 mya since latest submergence) makes it a suitable test for hypotheses. Its youth and remoteness necessitates long-distance immigration according to both EE and PT, providing a natural laboratory for an *experimentum crucis.* Recently, Kingston *et al.* (2003) provided range data for each of the 114 plant species of the Pitcairn group. If indeed it is true that "there is absolutely no reason why" plants and animals that have crossed the entire width of the Pacific should have a wide ranging distribution, then it should be likewise true for taxa that have managed half that distance. Accordingly, the Pitcairn group should exclusively share a significan percentage of taxa with South America (or Australia) that are limited in dispersal capability and unrelated to taxa found on any other ocean island group. (Indeed, given the frequency and extent of jump dispersal required by PT, perhaps the Pitcairn group may be disproportionately related to taxa exclusively found in a continental region >8000 km away, say, the east coast of North America).

As Kingston *et al.* (2003) write: "The flora of the Pitcairn Islands is derived from the flora of other island groups in the south-eastern Polynesian region, notably those of the Austral, Society, and Cook Islands. Species with a Pacific wide distribution dominate the overall Pitcairn group flora" They also note that Pitcairn, unlike New Zealand, did not exclusively share any plant with South America. Instead, plants that had dispersed between Pitcairn and South America also colonized other regions around the Pacific. According to their data set, every one of the non-endemic species of flora on Pitcairn also occur on some other Pacific island group — and 18 of the 19 endemics appear most closely related to taxa that occur on the Austral Islands as well as other islands around the Pacific The other endemic, an *Abutilon* species, appears akin to taxa from West Polynesia, Marquesas, Tuamotu, and Gambier Islands. None of the 114 species could be classified as poor-dispersing. None were sister to narrow-range plants that appear in only one other distant continental region.

An analysis of uncontroversial cross-ocean vertebrate rafting events over the past 30 mya reinforces this same point. The greatest trans-marine dispersers amongst the terrestrial vertebrates by far are the very tiny geckos and skinks, which may be the only vertebrates deserving the description "oceanic voyagers." *Mabuya* lizards (Carranza & Arnold, 2003) and *Tarentola* geckos (Carranza et al., 2000) have rafted considerable distances throughout the Atlantic, while Indian Ocean islands teem with *Phelsuma* geckos (Austin et al., 2004). Excluding these very small reptiles, no cross-ocean (i.e., entire ocean) rafting events of terrestrial vertebrates are even alleged to have occurred in the last 20 mya (McCarthy, 2005b). Moreover, the complicated and extensive distributions of each of these miniature reptiles once again confirm that taxa that can disperse across wide marine gaps tend to be extremely wide-ranging and occur on many oceanic islands.

This provides empirical confirmation for an otherwise obvious point — taxa that are able to colonize the remoter regions of the Pacific can annex nearer regions too. Often, the nearer regions are used as stepping stones. The fact that many of the trans-Pacific disjuncts cited (e.g., Matile, 1990; Heads, 1999; O'Foighil et al., 1999; McCarthy, 2003; Pregill & Worthy, 2003) are not found on any other oceanic island groups and are restricted to a very narrow range of regions, which also happen to be juxtaposed in EE paleomaps, adds another biogeographical implausibility to PT. Not only do the palpable biophysical restrictions of the referenced taxa appear incompatible with long-distance, trans-marine dispersal, their absence from other oceanic islands provides additional confirmation as to their sedentary nature. Apparently, these disjunct taxa, which have an easy time crossing wide oceanic barriers that are hypothetical, have a much more difficult time with narrow marine barriers that are real.

FURTIVE FOSSILS

Briggs responds to the disjunct terrestrial tetrapods shared by East Asia and South-western North America with the suggestion of range expansion across "a continuous, terrestrial habitat across the northern part of the globe from western North America to eastern Siberia". Figure 8.2 shows the plate tectonic reconstructions for 200 mya by Schettino & Scotese (2001). A few of the trans-oceanic regions that share a significant number of poor-dispersing, pre-Jurassic taxa (Ager, 1986; Briggs, 1987; Newton, 1988; Shields, 1998) and have a high index of similarity for Late Trias-sic–Early Jurassic tetrapods (Hotton, 1992) have been highlighted. Figure 8.3 also includes a reconstruction of the polar view and the northern connection to which Briggs refers. As is clear, southwestern North America is essentially on the opposite side of the planet from South China and Indochina, and the global route of range expansion is no less significant when taxa disregard climatic differentiations (Rieppel et al., 2000) and spread across the pole.

Also, many of the disjunct taxa are not found in Siberia, European Russia, Northern Europe, Western Europe, Greenland, or eastern North America (Hotton,

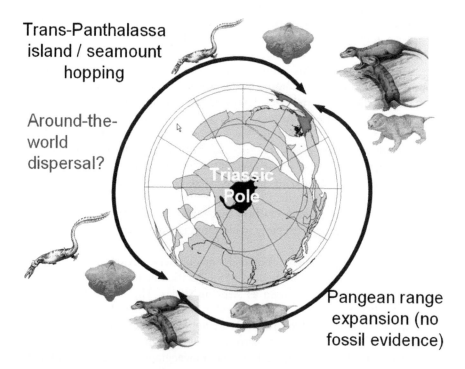

Trans-Panthalassa island / seamount hopping

Around-the-world dispersal?

Triassic Pole

Pangean range expansion (no fossil evidence)

FIGURE 8.3 A polar satellite ("top-of-the-world") view of the Early Jurassic plate tectonic paleomap of Figure 8.2 (from Schettino & Scotese, 2001). East Asia and Western North America (which is not in view) are placed on opposite sides of the globe. This paleomap requires that disjunct terrestrial and freshwater fauna and flora spread across the full width of Pangaea, despite lack of fossil evidence in all intervening regions, while disjunct shallow marine fauna island-hopped or seamount-hopped across the full breadth of Panthalassa.

1992; Shields, 1998). East Asia, in fact, shares more tetrapod families with North America than with any other region (Hotton, 1992), and Yunnan, China, and Arizona exclusively share two different sister taxa from the Tritylodontidae (Luo & Wu, 1995; Setoguchi et al., 1999). Currently, countless globally disjunct regions, like, say, Argentina and Turkey, are connected via a continuous stretch of continental material, but these regions do not share a significant percentage of poor-dispersing taxa found nowhere else in the world.

Similarly, Briggs dismissed all of the Late Triassic disjunctions of dozens of taxa described by Shields (1998) because "the organisms in question might have been distributed across the width of Pangea", but their fossils have yet to be unearthed in the intervening regions. Certainly, it is possible that many of the taxa occurred in regions where fossil evidence is currently lacking, but a basic principle of biogeography states that the greater number of fossil absences or extinctions required, the less parsimonious the theory. Simple statistics supports that notion. Briggs (1987) does not assume that taxa exclusively shared by Africa and South America in the Triassic spread there via undiscovered ancestors in Eurasia and North America. Instead, he, like almost all other modern biogeographers, agrees the similarity of the Triassic fauna leaves "no doubt" that South America and Africa were joined. Likewise, the reason Briggs (2003) does not contend that various Late Cretaceous Eurasian taxa reached India via Africa is because the taxa are not found in Africa. Clearly, biogeographers find it more parsimonious to accept that taxa probably did not occur in regions in which no evidence for them is found. Also, defenders of plate tectonics must not simply argue that a *few* of these Late Triassic terrestrial disjuncts may have occurred in *a few* of the inter-vening regions, they must contend that *every* disjunct taxon occurred in *every* intervening region.

Moreover, many of the Late Triassic disjuncts described by Shields (1998) are identical species. So defenders of plate tectonics not only have to maintain that the entire width of Pangaea provided no obstacles to dispersal, but that it also furnished no obstacles to gene fl w. It seems unlikely that these fossil taxa have escaped all detection across Pangaea despite being so thickly settled, so globally pervasive, that gene fl w was continuous from one end of the super-continent to the other.

Finally, even assuming that all these terrestrial taxa in the Late Triassic managed to elude the fossil hunters everywhere except the Pangaean margins, PT once again cannot escape trans-Panthalassa jaunts of poor dispersers. Some of the disjunct sister taxa are shallow marine tetrapods confined to near-shore habitats, and no marine corridor existed between western Tethys and western North America at that time (Ziegler, 1988; Dore, 1991; Ricou, 1996, Rieppel et al., 2000). As Rieppel *et al.* (2000) point out, the distribution of marine tetrapods requires, according to PT paleomaps, island-hopping across Panthalassa for both pistosaurs and thalattosaurs. Jiang *et al.* (2004) have recently described another trans-Panthalassa link between shallow marine reptiles: the Late Triassic *Xinpusaurus* from Guizhou, China, and its sister, the Californian *Nectosaurus*. The disjuncts also include a myriad of shallow-water and benthic marine fossil fauna, including foraminifers, rugose corals, scler-actinian corals, bivalves, gastropods, spongiomorphs, sponges, microcrinoids, ano-muran crustacean ichnofossils, and ammonoid cephalopods (e.g., Newton, 1988).

The PT theory entails the hypothesis that neither the full width of Panthalassa nor the full width of Pangaea provided an effective dispersal barrier to any type of taxon. Followers of PT must attribute the disjunctions to massive range expansions of a myriad of East Asian and Western American taxa in both directions around the planet, with all of the taxa meeting in the same location on the opposite side of the globe (Figure 8.3). Despite being distributed along an entire low- to mid-latitudinal circumference, these taxa only left discovered fossils in two regions — regions that lack any pre-Jurassic crust between them and happen to be juxtaposed in EE paleomaps.

DISPERSAL COUNTS, BIOTIC SIMILARITY, AND THE DISTANCE EFFECT

Briggs (2004) referenced Fallaw (1983) as an example of a biogeographical argument supporting a wider pre-Pacific superocean due to an increase in faunal similarity of marine invertebrates on opposite sides of the Pacific since the Early Cretaceous. As far as I am aware, Fallaw's paper (1983) is one of only two *biogeographical* analyses purported to favor the hypothesis of an ancient, oceanic separation of continents — Simpson's paper (1943) supporting continental stabilism being the other. Both analyses involve comparisons of Simpson's indices of similarity from Mesozoic or Early Cenozoic times with more recent indices. Both papers contain the same error. Neither Fallaw nor Simpson accounted for the fact that *observed* indices of similarity will significantly increase without a corresponding increase in the *actual* index of similarity as the time approaches the present due to the generally increasing quality of the fossil record or the superior knowledge of extant taxa. As the percentage of the fossil families (or genera or species) that are discovered increases, so does the ratio of the *observed* index of similarity with respect to the *actual* index of similarity. For example, if two regions have an actual index of similarity of 1 for tetrapod families in the Late Triassic, but only 50% of all tetrapod families are discovered in the primary region, then the *observed* index of similarity will likely be 0.5.

In the data used by Fallaw (1983), the number of discovered genera in each of the IndoPacific-Asia and American regions generally decreases with increasing age from >400 in the Pleistocene to <100 in the Pliensbachian, a stage of the Early Jurassic. Unless this accurately reflects an increase in diversity at the genus level since the Early Jurassic by more than a factor of four, the data imply a significant and general decrease in the quality of the fossil record as age increases. This entails a corresponding and misleading decrease in the *observed* indices of similarity over time due to basic statistical proclivities. Simpson's paper (1943), which argues for an ancient and wide Atlantic, contains this same fl w.

A more efficacious analysis based on the distance effect would have compared indices of similarity with those of the Atlantic during the same time frame, for the opening and expansion of the Atlantic is no longer the subject of debate. While Fallaw did not furnish such an analysis, Briggs (1987) has written that, "most of the old aquatic and terrestrial invertebrates and the most primitive plants [of South America] show a stronger affinity to Australia than to Africa", which is to say, the trans-Pacific affinities were greater than the trans-Atlantic ones.

Sanmartín and Ronquist (2004) have recently provided a comparison of dispersal events across the various southern oceans, seas, and channels in their wide-ranging study of 54 animal and 19 plant phylogenies (1393 terminals), making it "the largest biogeographical analysis of the Southern Hemisphere attempted so far". To avoid sampling bias, Sanmartín and Ronquist provided results in terms of a ratio of dispersal events "that we should have seen if all areas had been equally well represented in the data set". This allows comparison of dispersal counts across the Pacific with those across the Atlantic Ocean, the Indian Ocean, the Tasman Sea, the Torres Strait, and the Mozambique Channel. The results are unambiguous. The trans-Pacific pairs of regions that are juxta-posed in EE paleomaps (e.g., Shields, 1979) and fit together like pieces of a puzzle (Scalera, 1993; McCarthy, 2003), South America and New Zealand/ Australia, also happen to be the pairs of regions that have anomalously high dispersal frequencies and conspicuously flout the "distance effect" (MacArthur & Wilson, 1967) on PT paleomaps.

As noted by Sanmartín and Ronquist (2004), the oft-discovered sister-group rela-tionship and the best plant area cladogram (southern South America, New Zealand,) (Australia)) conflicts with conventional geological predictions for vicariance but is consistent with a "different sequence of vicariance events (Glasby & Alvarez, 1999)", which is to say, it matches EE theory. However, they felt that a "more convincing explanation is that it is the result of long-distance dispersal between New Zealand and South America after New Zealand drifted away from west Antarctica, 80 MYA". According to their data, the ratio of New Zealand to Southern South America dispersal events among both animal and plant taxa is greater than the ratio of dispersal events between every other pair of regions — except for two pairs indis-putably linked by vicariance: Australia–Southern South America and New Zealand–Australia. (Sanmartín & Ronquist [2004] attributed the Australia–Southern South America distributions to range expansion through Antarctica, though they did note "none of the studied groups are now present there.") Incredibly, a higher ratio of animals traveled from New Zealand to South America than in either direction between Madagascar and Africa, New Guinea and Australia, New Caledonia and New Zealand, India and New Guinea, and even Northern South America and South-ern South America. Plant taxa also showed a much greater propensity to cross the Pacific than the narrower, juvenile Atlantic Ocean, Indian Ocean, the Torres Strait, and even the Mozambique Channel. Wind and currents cannot be credited, for the West Wind Drift (i.e., "the prevailing direction of winds and ocean currents") opposes the dominant direction of South America–New Zealand plant dispersal. Removing the hypothesis of a now-subducted superocean, crustal age data of the southern Pacific demand that South America remained in direct contact with Australia/Lord Howe Rise and New Zealand into the Cretaceous and remained in proximity to New Zealand (separated by a narrow stretch of South Pacific crust) into the Early Tertiary (Figure 8.1c). As the figures from Sanmartín and Ronquist (2004) show, a dispro-portionately high number of dispersal events are between these trans-Pacific regions.

BRIEF RESPONSES

Briggs' explanation for the disjunctions of marine taxa by highlighting other taxa of high vagility that have crossed the expansive and deep East Pacific Barrier neglects two significant facts. First, the taxa cited do not occur in the East Pacific; they only inhabit the Caribbean and the Indo—West Pacific (Anderson, 2000; Mejia et al., 2001.) Like their terrestrial counterparts, these marine disjuncts reside over ancient crustal regions on opposite sides of the Pacific (the Caribbean and the West Pacific but have managed to bypass the region that has only recently formed between them (the entire East Pacific) Second, the taxa were poor dispersers. For example, the loliginid squid lack teleplanic larvae and exclusively inhabit near coastal regions (Anderson, 2000).

Briggs refers to the trans-Pacific shallow marine fauna of the Triassic and Jurassic as occurring on displaced terranes in Western North America. However, as noted:

1. Many Tethyan marine fossils are found on American craton-related regions, not allegedly displaced terranes (Newton, 1988).
2. Some of the terranes include a mixture of North American and Tethyan elements (Newton, 1988).
3. Some of the migrations appear to have been from east to west (Ager, 1986), implied, for example, by Japanese regions that have North American elements (Tazawa et al., 1998).
4. Hallam (1986) acknowledged that, even in plate tectonics, the notion of significant east-west motion of the terranes with respect to North America has now been discounted, except perhaps in Cache Creek.

Briggs wrote: "The final part of the paper is devoted to a discussion that contains some naive observations: (1) if everyone agrees that the Atlantic and Indian Oceans were closed in pre-Jurassic times, why could not the Pacific Ocean also have been closed?" (Briggs, 2004, pp. 856–857).

Certainly, that would have been naive had it been the argument. Instead, the point was that all of the significant and oft-cited evidence that supports a closed Atlantic also occurs in the Pacific This includes:

1. A myriad of trans-oceanic fossil links of poor-dispersing taxa
2. Matching geological outlines
3. Mid-ocean spreading ridges between the continents
4. The absence of pre-Jurassic ocean crust and pre-Jurassic ocean fossils between the regions
5. The isotopic and temporal similarity of a large igneous province on opposite sides of the ocean (i.e., the Ontong Java and Caribbean plateaus) (Arndt & Weis, 2002).

One cannot well argue that this evidence is only significant when it occurs in the Atlantic but not the Pacific

GEOLOGICAL CONCERNS

Most of Briggs objections to the EE view were geological in nature. While all of his points cannot be answered in detail in a single paper, it is important to stress that geophysicists and geologists were the first to conclude the Earth was expanding (e.g., Bruce Heezen, Warren Carey, Otto Hilgenberg), and geophysicists and geologists mainly comprise the followers today (e.g., Giancarlo Scalera, James Maxlow, David Ford). EE is not a theory with a biogeographical pedigree. It is, first and foremost, a geological theory, rich in history, that, unlike continental drift, predicted seafloor spreading and the juvenile nature of all ocean basin crust (Hilgenberg, 1933 — also discussed in Carey, 1988; Scalera, 1997). While most EE researchers are aware of the distributional evidence, most books on EE brim with paleomagnetic, geological, sedimentary, geodetic, and other forms of geophysical evidence (e.g., Carey, 1988; Scalera & Jacob, 2003). In fact, Carey (1988), Maxlow (2001), Scalera (2003), and McCarthy (2005a) have answered many of the questions raised by Briggs (2004) and have underscored other geological facts that contradict fundamental PT assumptions.

For example, Briggs concern about "the absence of a drastic fall in sea level since the Triassic that would have been caused by expansion", presumes that the volume of surface water has remained stable for hundreds of millions of years. However, tidal gauge data observed throughout the twentieth century (Miller & Douglas, 2004) and more recent satellite measurements (Cabenas et al., 2001) have shown global sea level has been rising 1.5–3.2 mm yr^{-1}. Until recently researchers had attributed global seal level rise (GSLR) to temperature- and salinity-related changes in volume, but in 2004, Miller and Douglas showed that such volume changes can "account for only a fraction of sea level change, and that mass change plays a dominant role in twentieth-century GSLR" (Miller and Douglas, 2004, p. 408). The most parsimonious explanation in EE is that GSLR is just the continuation of the process of ocean formation — most likely linked to outfl w from hydrothermal vents (McCarthy, 2005a).

Certainly, all geological evidence must be carefully considered when fashioning distributional explanations. But as Derek Ager pointed out in his 1986 presidential address to the *British Association for the Advancement of Science* (Ager, 1986), the debate today, like the debate in the first part of the twentieth century, is really between geophysical *assumptions* and biogeographical *facts*. Ager understood that the recent revolution in geology began not with mainstream geologists and geophysicists but with scientific outsiders who adhered to the central realities of biogeography. In the introduction to his address, Ager noted that in his time geophysicists were claiming that mobile continents were impossible, and now "they seem to take all the glory for it as if it were their idea in the first place" (Ager, 1986, p. 377). Ager went on to provide a thorough analysis of brachiopod disjunctions across the Atlantic, Indian, and Pacific, distributions that imply an expanding Earth. In the conclusion, Ager states that, "I am not deterred by geophysicists telling me that [expanding Earth] is impossible. I have heard that sort of story before from the geophysicists" (Ager, 1986, p. 388). Ager's point is hard to dismiss. The reason why, in the first half of the twentieth century, followers of biogeography had bested mainstream geologists in their very own subject is still true today — while we *speculate* about the formation and inner-workings of planets, we *know* which taxa can cross oceans.

SUMMARY

The case for a closed Pacific entailed by the trans-Pacific distributions supersedes the typical vicariance-dispersal debate that strains the presses at Blackwell Publishing. Regardless of the battle between various biogeographical factions, practically all researchers agree on a few basic distributional realities:

1. Ocean crustal age data require trans-marine dispersal as almost the exclusive mechanism of colonization of many juvenile, volcanic oceanic islands.
2. Both the high level of endemism and the dearth of certain kinds of taxa on remoter oceanic islands confirm the importance of the *distance effect* and the insuperable obstacle that marine gaps present to many kinds of plants and animals.
3. Taxa that can and do cross extensive seawater barriers tend to be wide ranging, often occurring on other accessible oceanic islands.
4. Disjunctions of an enormous variety of poor-dispersing terrestrial taxa by the full breadth of an ocean are not most reasonably explained by cross-ocean jaunts and range expansion and fossil absences (du Toit's Principle).

These last few widely-agreed upon facts are what confront the PT hypotheses, and they are not controversial.

Moreover, as noted, at no point has anyone challenged these principles. No one has claimed that PT requirements of cross-ocean rafting, cross-ocean island hopping, or convenient fossil absences are biogeographically preferable. Clearly, the hundreds of taxa cited by trans-Pacific researchers are more apt to access proximal regions rather than the other side of the world, so the question of which theory provides the more reasonable distributional explanation cannot be debated.

It is telling to note that, after accepting the vicariant explanations for trans-Atlantic and trans-Indian Ocean distributions, we now find biogeographical controversy due to linkage of pairs of regions like New Zealand and Southern Chile (Heads, 1999) or Fiji and the Neotropics (Matile, 1990; Pregill & Worthy, 2003) or the Caribbean and West Pacific (Glasby, 1999; Mejia et al., 2001; Anderson, 2000) or China and Arizona (in the Late Triassic) (Luo & Wu, 1995; Shields, 1998; Setoguchi et al., 1999) or India and Eurasia (Late Cretaceous) (Briggs, 2003) — that is, the only remaining regions in the world where geologists are still trying to hypothesize extreme Mesozoic separation by ancient (and vanished) seafloo . Certainly, the distribution of a particular taxon may be fragmented, even globally so, but we never seem to find similar disjunctions among a multiplicity of poor-dispersing taxa repeatedly connecting other widely separated territories. We, for example, do not fin researchers fretting over the "mystery of the Wisconsin–Ceylon disjunctions" or the "problematic Western Australia–Western France distributions". Yet, wind, phoresy, convenient fossil absences, etc., should also conspire to construct misleading and seemingly endless biotic links between other distant lands — if, that is, the notion is plausible.

Similarly, wind, currents, and rafts should also scheme to ferry poor dispersers from one particular continental region to another particular oceanic island group >8000 km away (or, at the very least, >2500 km away), while avoiding all other islands and regions significantly closer. Yet, all the truly wondrous biotic distributions, all the examples of directed, limited-range, continent–island distributions exclusively favor primitive taxa from western South America and pre-Oligocene West Pacific islands, like Fiji, New Caledonia, and New Zealand. It cannot be coincidental that islands devoid of the geologically controversial are islands devoid of the biogeographically marvelous. Pitcairn, like Hawaii, cannot boast any miraculous vertebrate rafting trips of narrow-range taxa. And all of Pitcairn's species of flora have very obvious means of long-distance dispersal and come from wide ranging ancestors that occur on other oceanic island groups (Kingston et al., 2003). The relevant geological difference is that the Pitcairn group formed in the middle of a large ocean basin that still exists — so its significant separation from South America (and Australia) is not theoretical. Perhaps this is why all the expected biogeographical consequences of marine remoteness, particularly from South America, are on display in Pitcairn — but are not evident among many of the primitive taxa from the more ancient islands like New Zealand, New Caledonia, or Fiji, despite being nearly twice as far away.

The outstanding biogeographical problems confronting PT are as follows:

1. No researcher supporting PT has provided a biogeographical justificatio for the difference in conclusions inferred from the pre-Jurassic, trans-Atlantic and the pre-Jurassic, trans-Pacific disjunctions. If the trans-Atlantic disjunctions of pre-Jurassic fossil tetrapods, flora, shallow marine fauna, etc., provide evidence for a closed Atlantic, then a greater quantity of trans-Pacific disjunctions of the same type of taxa from the same time period provide evidence for a closed Pacific

2. No researcher supporting PT has been able to describe other disjunctions analogous to the Late Triassic, trans-Pacific disjunctions that are not currently explained by vicariance. No other pair of regions meet the following criteria:
   a. The pair *exclusively* share a wide variety of poor-dispersing taxa (e.g., terrestrial tetrapods, freshwater fish) at some point in the last 300 mya.
   b. They are separated by a wide marine barrier (>2500 km) and/or a terrestrial continental route >6000 km throughout the existence of the taxa.
   c. They display a relatively high index of similarity for non-volant, terrestrial taxa and are separated by at least one interim region that is fossiliferous.
   d. The disjunctions are not explained by vicariance.

3. No researcher supporting PT has been able to highlight other trans-oceanic distributional relationships involving island taxa analogous to the southern trans-Pacific disjuncts shared by South America and New Zealand, New Caledonia or Fiji. No other island group exclusively shares a myriad of other poor-dispersing, narrow-range taxa with a single source continental region >8000 km (or >5000 km distant or even > 2500 km) distant — when other suitable regions and islands are nearby.

The biogeographical problems this presents to PT appear to be insurmountable, and it is difficult to foresee how a defender of orthodoxy would respond to them.

How, for example, could a researcher argue that the trans-Atlantic fossil disjunctions are more persuasive than the trans-Pacific ones? Anyone who notes the classic textbook examples of the trans-Atlantic affinities of *Mesosaurus* and *Cynognathus* must then discuss trans-Pacific affinities of *Corosaurus* (Rieppel, 1999), *Dinnebitodon* and *Kayentatherium* (Luo & Wu, 1995; Setoguchi et al., 1999). If a biogeographer cites the high index of similarity of tetrapod families in the Late Triassic–Early Jurassic between South America and South Africa (64), he must confront the fact that East Asia–North America boast the exact same index of similarity (64) (Hotton, 1992). Those who point to the brachiopods shared by eastern North America and Western Europe (Ager, 1986) must then explain all the brachiopods shared by western North America and East Asia (also Ager, 1986). Proponents of phylogenetics and cladograms will find a greater number linking the margins of the Pacific than the Atlantic. Once trans-Atlantic distributions run empty, researchers have to continue to explain a myriad of trans-Pacific distributions involving freshwater fish, insects, flora, and non-aquatic vertebrates (e.g., Matile, 1990; Shields, 1998; Heads, 1999; Pregill & Worthy, 2003). Moreover, the counter-argument that the trans-Pacific disjunctions are the result of range expansion and fossil absences in the interim regions applies more plausibly to taxa shared by South Africa and South America than to taxa on the opposite margins of Pangea. Likewise, hypotheses of cross-ocean rafting, island hopping, or seamount hopping are much easier to suppose across the Atlantic than across a superocean three to four times wider. Every type of biogeographical evidence on behalf of a closed Atlantic can be matched and bested in the Pacific, and every current rationale used to dismiss the trans-Pacific links are more fitting for their trans-Atlantic counterparts — and, indeed, have already been used in a failed effort to conceal their biogeographical significance (Simpson, 1940, 1943).

This is no small point. For the evidence provided by the trans-Atlantic (and trans-Indian) fossil disjunctions are so basic and compelling that primers and popular works on geology usually introduce the concept of Pangea to students and interested laymen alike by highlighting the pre-Jurassic trans-oceanic distributions of poor-dispersers (e.g., Erickson, 1992; Kious & Tilling, 1996). Few, if any, scientists have attempted to challenge the validity of this reasoning in the last few decades. Instead, Simpson-like efforts to evade the obviousness of this argument are discussed only in unflattering, historical autopsies of the stabilist paradigm (e.g., Nelson & Ladiges, 2001). Such analyses help reveal how mainstream prejudices often lead to immoderate rationalizations. Quite simply, these distributional hypotheses, which have repeatedly been invoked to resuscitate geological tradition, whether involving continental stabilism or PT, do not belong to a coherent approach to modern biogeography; they belong to an infamous moment in the history of science.

ACKNOWLEDGEMENTS

I would like to thank Malte Ebach, Michael Heads, and John Grehan for their never-ending patience and help with material and publication. The significance of their generous support and encouragement cannot be overstated. I also greatly appreciate the online map-making tools of Antonio Schettino and Christopher R. Scotese for generating clear and beautiful plate tectonic reconstructions.

REFERENCES

Ager, D.V. (1986). Migrating fossils, moving plates, and an expanding earth. *Modern Geology*, 10, 377–390.

Anderson, F.E., (2000). Phylogeny and historical biogeography of the loliginid squids (Mollusca: Cephalopoda) based on mitochondrial DNA sequence data. *Molecular Phylogenetics and Evolution*, 15, 191–214.

Arndt, N. & Weis, D. (2002). Oceanic plateaus as windows to the Earth's Interior: an ODP Success story. *Joides Journal, Special Issue*, 28, 79–84.

Austin, J.J., Arnold, E.N. & Jones C.G. (2004). Reconstructing an island radiation using ancient and recent DNA: the extinct and living day geckos (*Phelsuma*) of the Mascarene islands. *Molecular Phylogenetics and Evolution*, 31, 109–122.

Briggs, J.C. (1987). *Biogeography and Plate Tectonics*. Elsevier, Amsterdam.

Briggs, J.C. (2003). The biogeographic and tectonic history of India. *Journal of Biogeography*, 30, 381–388.

Briggs, J.C. (2004). The ultimate expanding earth hypothesis. *Journal of Biogeography*, 31, 855–857.

Brundin, L. (1966). Transantarctic relationships and their significance, as evidenced by chironomid midges with a monograph of the subfamilies Podonominae and Aphroteniinae and the austral Heptagyiae. *Kunglica Svenska Vetenskapsakademiens Handlingar*, 11, 1–472 + 30 plates.

Cabanes, C., Cazenave, A. & Le Provost, C. (2001). Sea level rise during past 40 years determined from satellite and in situ observations. *Science*, 294, 840–842.

Carey, S.W. (1988). *Theories of the Earth and Universe. A history of Dogma in Earth Sciences*. Stanford University Press, Stanford, CA.

Carranza S. & Arnold, E.N. (2003). Investigating the origin of transoceanic distributions: mtDNA shows *Mabuya* lizards (Reptilia, Scincidae) crossed the Atlantic twice. *Systematics and Biodiversity*, 1, 275–282

Carranza S, Arnold E.N., Mateo J.A. & López-Jurado L.F. (2000). Long-distance colonization and radiation in gekkonid lizards, *Tarentola* (Reptilia: Gekkonidae), revealed by mitochondrial DNA sequences. *Proceedings of the Royal Society of London B*, 267, 637–649.

Cogger, H.G. (1974). Voyage of the banded iguana. *Australian Natural History*, 18, 144–149.

Croizat, L. (1958). *Panbiogeography*, pp. xii + 1731. Published by author in three volumes, Caracas.

Cranston, P.S. (2003). Biogeographic patterns in the evolution of Diptera. In *The Evolutionary Biology of Flies* (ed. by Wiegmann, B.M. & Yeates, D.K.). Columbia University Press, New York.

Craw, R.C., Grehan, J.R. & Heads, M.J. (1999). *Panbiogeography — Tracking the History of Life*. Oxford University Press, Oxford, UK.

Diamond, J.M. (1990). New Zealand as an archipelago: an international perspective. In *Ecological Restoration of New Zealand Islands*. (ed. by Towns, D.R., Daugherty, C.H. & Atkinson, I.A.E.), pp. 3–8. Department of Conservation, Wellington, New Zealand.

Dore, A.G. (1991). The structural foundation and evolution of Mesozoic seaways between Europe and the Arctic. *Palaeogeography, Palaeoclimatology, Palaeoecology*, 87, 441–492.

du Toit, A. (1944). Tertiary mammals and continental drift. A rejoinder to George G. Simpson. *American Journal of Science, 242, 145–163*.

Erickson, J. (1992). *Plate Tectonics: Unraveling the Mysteries of the Earth*. Facts on File, Inc. New York, pp. 8–9.

Fallaw, W.C. (1983). Trans-Pacific faunal similarities among Mesozoic and Cenozoic inverte-brates related to plate tectonic processes. *American Journal of Science,* 283, 166–172.

Gibbons, J.R.H. (1981). The biogeography of *Brachylophus* (Iguanidae) including description of a new species, *B. vitiensis,* from Fiji. *Journal of Herpetology,* 15, 255–273.

Glasby, C.J. (1999). The Namanereidinae (Polychaeta; Nereididae), Part 2, cladistic bioge-ography. *Records of the Australian Museum,* Supplement 25, 131–144.

Glasby, C.J. & Alvarez, B. (1999). Distribution patterns and biogeographic analysis of Austral Polychaeta (Annelida). *Journal of Biogeography,* 26, 507–533.

Grehan, J.R. (1991). Panbiogeography 1980–1990: Development of an earth/life synthesis. *Progress in Physical Geography,* 15, 331–363.

Hallam, A. (1986). Evidence of displaced terranes from Permian and Jurassic faunas around the Pacific margins. *Journal of the Geological Society, London,* 143, 209–216.

Heads, M. (1999). Vicariance biogeography and terrane tectonics in the South Pacific; analysis of the genus *Abrotanella* (Compositae). *Biological Journal of the Linnean Society,* 67, 391–432.

Hennig, W. (1960). Die Dipteren-Fauna von Neuseeland als systematisches und tiergeogra-phisches Problem. *Beitrage zur Entomologie,* 10, 221–329.

Hilgenberg, O.C. (1933). *Vom Wachsenden Erdball (On the Growing Earth).* Published by the author, Selbstverlag, Berlin.

Howarth, F.G., Nishida, G. & Asquith, A. (1995). Insects of Hawaii. In *Our Living Resources: A Report to the Nation on the Distribution, Abundance, and Health of U.S. Plants, Animals, and Ecosystems* (ed. by LaRoe, E.T., Farris, G.S., Puckett, C.E., Doran, P.D. & Mac, M.J.). U.S. Department of the Interior. National Biological Service, Washington, D.C. Available at http://biology.usgs.gov/s+t/noframe/t068.htm.

Hotton, N. (1992). Global distribution of terrestrial and aquatic tetrapods, and its relevance to the position of continental masses. In *New Concepts in Global Tectonics.* (ed. by Chatterjee, S. & Hotton, N., III), pp. 267–285. Texas Tech University Press, Lubbock.

Humphries, C.J. & Parenti, L. R. (1999). *Cladistic Biogeography / Second Edition: Interpret-ing Patterns of Plant and Animal Distributions* pp. 36, 134, 143–144, 150. Oxford University Press, Oxford.

Jacobi, J.D. (2001). Overview and conservation status of Hawaiian ecosystems. Oral presen-tation at Annual Meeting of the Society for Conservation Biology, 29 July–1 August, 2001. Hilo, Hawaii.

Jiang, D.-Y., Maisch, M.W., Sun, Y.-L., Matzke, A.T. & Hao, W.-C., (2004). A new species of *Xinpusaurus* (Thalattosauria) from the Upper Triassic of China. *Journal of Verte-brate Paleontology,* 24, 80–88.

Kingston, N., Waldren, S. & Bradley, U. (2003). The phytogeographical affinities of the Pitcairn Islands—a model for south-eastern Polynesia? *Journal of Biogeography,* 30, 1311–1328.

Kious, W.J. & Tilling R. (1996). *This Dynamic Earth, The Story of Plate Tectonics.* United States Government Printing Office. Washington D.C. Available online at: http://pubs.usgs.gov/publications/text/dynamic.html.

Loope, L.L. (1998). Hawaii and the Pacific Islands. In *Status and Trends of the Nation's Biological Resources, Volume 2* (ed. by Mac, M.J., Opler, P.A., Puckett Haecker, C.E. & Doran, P. D.), pp. 747–774. United States Department of the Interior. United States Geological Survey, Reston, Virginia. Available at http://biology.usgs.gov/s+t/SNT/noframe/pi179.htm.

Luo, Z. & Wu, X.-C. (1995). Correlation of vertebrate assemblage of the lower Lufeng Forma-tion, Yunnan, China. In *Sixth Symposium on Mesozoic Terrestrial Ecosystems and Biota, Short Papers* (ed. by Sun, A. & Wang, Y.), pp. 83–88. China Ocean Press, Beijing.

MacArthur, R.H. & Wilson, E.O. (1967). *The Theory of Island Biogeography.* Princeton University Press, Princeton.

Matile, L. (1990). Recherches sur la systématique et l'évolution des Keroplatidae (Diptera, Mycetophiloidea*). Mémoirs Museum national d'Histoire naturelle* série A, 148, 1–682.

Maxlow, J. (2001). Quantification of an Archaean to Recent Earth Expansion Process Using Global, Geological and Geophysical Data Sets. Ph.D. thesis, Curtin University of Technology. http://adt.curtin.edu.au/theses/available/adt-WCU20020117.145715/.

McCarthy, D. (2003). The trans-Pacific zipper effect: disjunct sister taxa and matching geological outlines that link the Pacific margins. *Journal of Biogeography,* 30, 1545–1561.

McCarthy, D. (2005a). Biogeographical and geological evidence for a smaller, completely-enclosed Pacific Basin in the Late Cretaceous. *Journal of Biogeography,* 32, 2161–2177.

McCarthy, D. (2005b). Biogeography and scientific revolutions. *The Systematist,* 25, 3–12.

McDowall, R. M. (2004). What biogeography is: a place for process. *Journal of Biogeography,* 31, 345–351.

Mejia, L. S., Acero, A., Roa, A. & Saavedra, L. (2001). Review of the fishes of the genus *Synagrops* from the Tropical Western Atlantic (Perciformes: Acropomatidae). *Caribbean Journal of Science,* 37, 202–209.

Miller, L. & Douglas, B.C. (2004). Mass and volume contributions to twentieth-century global sea level rise. *Nature,* 428, 406–409.

Mantovani, R. (1909). L'Antarctide. *Je m'instruis. La Science pour tous, n (degrees),* 38, 595–597.

Mueller, R.D., Roest, W.R., Royer, J.-Y., Gahagan, L.M. & Sclater, J.G. (1993). A digital age map of the ocean floo . SIO Reference Series 93–30. Scripps Institution of Oceanography. http://www.ngdc.noaa.gov/mgg/fliers/96mgg04.html

Nelson, G. & Ladiges, P.Y. (2001). Gondwana, vicariance biogeography and the New York School revisited. *Australian Journal of Botany,* 49, 389–409.

Newton, C.R. (1988). Significance of "Tethyan" fossils in the American Cordillera. *Science,* 242, 385–391.

O'Foighil, D.O., Marshall, B.A., Hilbish, T.J. & Pino, M.A. (1999). Trans-Pacific range extension by rafting is inferred for the flat oyster *Ostrea chilensis. Biological Bulletin,* 196, 122–126.

Pregill, G.K. & Worthy, T.H. (2003). A new iguanid lizard (Squamata, Iguanidae) from the Late Quaternary of Fiji, Southwest Pacific. *Herpetologica,* 59, 57–67.

Ricou, L.E. (1996). The plate tectonic history of the past Tethys ocean. In *The Ocean Basins and Margins* (ed. by Nairn, A.E.M., Ricou, L.-E., Vrielynck, B. & Dercourt, J.), vol. 8, pp. 3–70. Plenum Press, New York.

Rieppel, O. (1999). The sauropterygian genera *Chinchenia, Kwangsisaurus,* and *Sanchiaosaurus* from the Lower and Middle Triassic of China. *Journal of Vertebrate Paleontology,* 19, 321–337.

Rieppel, O., Liu, J. & Bucher, H. (2000). The first record of a thalattosaur reptile from the Late Triassic of southern China (Guizhou Province, PR China). *Journal of Vertebrate Paleontology,* 20, 507–514.

Sanmartín, I. & Ronquist, F (2004). Southern hemisphere biogeography inferred by event-based models: plant versus animal patterns. *Systematic Biology,* 53, 216–243.

Scalera, G.C. (1993). Non-chaotic emplacements of trench arc zones in the Pacific Hemisphere. *Annali de Geophisica,* 36, 47–53.

Scalera, G. (1997). Roberto Mantovani, the father of the continental drift and planetary dilatation. In *Geomagnetism and Aeronomy (with Special Historical Case Studies)* (ed. by Schroder, W.), pp. 108–110. Science Edition, Interdivisional Commission on History of IAGA (International Association of Geomagnetism and Aeronomy) and History Commission of Deutsche Geologische Gesellschaft.

Scalera, G. (2003). The expanding Earth: a sound idea for the new millennium. In *Why Expanding Earth? A Book in Honour of Otto Christoph Hilgenberg.* (ed. by Scalera, G. & Jacob, K.H.), pp. 181–232. Istituto Nazionale di Geofisica e Vulcanologia, Rome.

Scalera, G. & Jacob, K.H. (2003). *Why Expanding Earth? A Book in Honour of Otto Christoph Hilgenberg.* Istituto Nazionale di Geofisica e Vulcanologia, Rome.

Schettino, A. & Scotese, C.R. (2001). New Internet software aids paleomagnetic analysis and plate tectonic reconstructions. *Eos, Transactions, American Geophysical Union, Electronic Supplement,* 82. http://www.agu.org/eos_elec/010181e.html.

Sequeira, A.S. & Farrell, B.D. (2001). Evolutionary origins of Gondwanan interactions: how old are *Araucaria* beetle herbivores? *Biological Journal of the Linnean Society,* 74, 459–474.

Setoguchi, T., Matsuoka, H. & Matsuda, M. (1999). New discovery of an early Cretaceous tritylodontid (Reptilia,Therapsida) from Japan and the phylogenetic reconstruction of Tritylodontidae based on the dental characters. In *Proceedings of the Seventh Annual Meeting of the Chinese Society of Vertebrate Paleontology* (ed. by Yuanqing, W. & Tao, D.), pp. 117–124. China Ocean Press, Beijing.

Shields, O. (1979). Evidence for initial opening of the Pacific Ocean in the Jurassic. *Palaeogeography, Palaeoclimatology, Palaeoecology,* 26, 181–220.

Shields, O. (1996). Geologic Significance of land organisms that crossed over the Eastern Tethys "Barrier" during the Permo-Triassic. *Palaeobotanist,* 43, 85–95.

Shields, O. (1998). Upper Triassic Pacific vicariance as a test of geological theories. *Journal of Biogeography,* 25, 203–211.

Simpson, G.G. (1940). Antarctica as a faunal migration route. *Proceedings of the Sixth Pacific Science Congress of the Pacific Science Association,* 2, 755–768. *University of California Press, Berkeley & Los Angeles.*

Simpson, G.G. (1943). Mammals and the nature of continents. *American Journal of Science,* 241, 1–31.

Sites, J.R., Davis, S.K., Guerra, T., Iverson, J.B. & Snell. H.L. (1996). Character congruence and phylogenetic signal in molecular and morphological data sets: a case study in the living iguanas (Squamata, Iguanidae). *Molecular Biology and Evolution,* 13, 1087–1105.

Tazawa J., Ono T. & Hori, M. (1998). Two Permian lyttonid brachiopods from Akasaka, central Japan. *Paleontological Research,* 2, 239–245.

Waters, J.M., Lopez, J.A. & Wallis, G.P. (2000). Molecular phylogenetics and biogeography of galaxiid fishes (Osteichthyes: Galaxiidae): dispersal, vicariance, and the position of *Lepidogalaxias salamandroides. Systematic Biology,* 49, 777–795.

Ziegler, P.A. (1988). Evolution of the Arctic-North Atlantic and the western Tethys. *American Association of Petroleum Geologists Memoir,* 43, 1–198.

Index

Systematics Association Publications

1. Bibliography of Key Works for the Identification of the British Fauna and Flora, 3rd edition (1967)[†]
   Edited by G.J. Kerrich, R.D. Meikie and N. Tebble
2. Function and Taxonomic Importance (1959)[†]
   Edited by A.J. Cain
3. The Species Concept in Palaeontology (1956)[†]
   Edited by P.C. Sylvester-Bradley
4. Taxonomy and Geography (1962)[†]
   Edited by D. Nichols
5. Speciation in the Sea (1963)[†]
   Edited by J.P. Harding and N. Tebble
6. Phenetic and Phylogenetic Classification (1964)[†]
   Edited by V.H. Heywood and J. McNeill
7. Aspects of Tethyan biogeography (1967)[†]
   Edited by C.G. Adams and D.V. Ager
8. The Soil Ecosystem (1969)[†]
   Edited by H. Sheals
9. Organisms and Continents through Time (1973)[†]
   Edited by N.F. Hughes
10. Cladistics: A Practical Course in Systematics (1992)[*]
    P.L. Forey, C.J. Humphries, I.J. Kitching, R.W. Scotland, D.J. Siebert and D.M. Williams
11. Cladistics: The Theory and Practice of Parsimony Analysis (2nd edition) (1998)[*]
    I.J. Kitching, P.L. Forey, C.J. Humphries and D.M. Williams

[*] Published by Oxford University Press for the Systematics Association
[†] Published by the Association (out of print)

SYSTEMATICS ASSOCIATION SPECIAL VOLUMES

1. The New Systematics (1940)
   Edited by J.S. Huxley (reprinted 1971)
2. Chemotaxonomy and Serotaxonomy (1968)*
   Edited by J.C. Hawkes
3. Data Processing in Biology and Geology (1971)*
   Edited by J.L. Cutbill
4. Scanning Electron Microscopy (1971)*
   Edited by V.H. Heywood
5. Taxonomy and Ecology (1973)*
   Edited by V.H. Heywood

6. The Changing Flora and Fauna of Britain (1974)*
   Edited by D.L. Hawksworth

7. Biological Identification with Computers (1975)*
   Edited by R.J. Pankhurst

8. Lichenology: Progress and Problems (1976)*
   Edited by D.H. Brown, D.L. Hawksworth and R.H. Bailey

9. Key Works to the Fauna and Flora of the British Isles and Northwestern Europe, 4th edition (1978)*
   Edited by G.J. Kerrich, D.L. Hawksworth and R.W. Sims

10. Modern Approaches to the Taxonomy of Red and Brown Algae (1978)
    Edited by D.E.G. Irvine and J.H. Price

11. Biology and Systematics of Colonial Organisms (1979)*
    Edited by C. Larwood and B.R. Rosen

12. The Origin of Major Invertebrate Groups (1979)*
    Edited by M.R. House

13. Advances in Bryozoology (1979)*
    Edited by G.P. Larwood and M.B. Abbott

14. Bryophyte Systematics (1979)*
    Edited by G.C.S. Clarke and J.G. Duckett

15. The Terrestrial Environment and the Origin of Land Vertebrates (1980)
    Edited by A.L. Pachen

16 Chemosystematics: Principles and Practice (1980)*
    Edited by F.A. Bisby, J.G. Vaughan and C.A. Wright

17. The Shore Environment: Methods and Ecosystems (2 volumes)(1980)*
    Edited by J.H. Price, D.E.C. Irvine and W.F. Farnham

18. The Ammonoidea (1981)*
    Edited by M.R. House and J.R. Senior

19. Biosystematics of Social Insects (1981)*
    Edited by P.E. House and J.-L. Clement

20. Genome Evolution (1982)*
    Edited by G.A. Dover and R.B. Flavell

21. Problems of Phylogenetic Reconstruction (1982)
    Edited by K.A. Joysey and A.E. Friday

22. Concepts in Nematode Systematics (1983)*
    Edited by A.R. Stone, H.M. Platt and L.F. Khalil

23. Evolution, Time And Space: The Emergence of the Biosphere (1983)*
    Edited by R.W. Sims, J.H. Price and P.E.S. Whalley

24. Protein Polymorphism: Adaptive and Taxonomic Significance (1983)*
    Edited by G.S. Oxford and D. Rollinson

25. Current Concepts in Plant Taxonomy (1983)*
    Edited by V.H. Heywood and D.M. Moore

26. Databases in Systematics (1984)*
    Edited by R. Allkin and F.A. Bisby

27. Systematics of the Green Algae (1984)*
    Edited by D.E.G. Irvine and D.M. John

50. Systematics and Conservation Evaluation (1994)‡
    Edited by P.L. Forey, C.J. Humphries and R.I. Vane-Wright

51. The Haptophyte Algae (1994)‡
    Edited by J.C. Green and B.S.C. Leadbeater

52. Models in Phylogeny Reconstruction (1994)‡
    Edited by R. Scotland, D.I. Siebert and D.M. Williams

53. The Ecology of Agricultural Pests: Biochemical Approaches (1996)**
    Edited by W.O.C. Symondson and J.E. Liddell

54. Species: the Units of Diversity (1997)**
    Edited by M.F. Claridge, H.A. Dawah and M.R. Wilson

55. Arthropod Relationships (1998)**
    Edited by R.A. Fortey and R.H. Thomas

56. Evolutionary Relationships among Protozoa (1998)**
    Edited by G.H. Coombs, K. Vickerman, M.A. Sleigh and A. Warren

57. Molecular Systematics and Plant Evolution (1999)
    Edited by P.M. Hollingsworth, R.M. Bateman and R.J. Gornall

58. Homology and Systematics (2000)
    Edited by R. Scotland and R.T. Pennington

59. The Flagellates: Unity, Diversity and Evolution (2000)
    Edited by B.S.C. Leadbeater and J.C. Green

60. Interrelationships of the Platyhelminthes (2001)
    Edited by D.T.J. Littlewood and R.A. Bray

61. Major Events in Early Vertebrate Evolution (2001)
    Edited by P.E. Ahlberg

62. The Changing Wildlife of Great Britain and Ireland (2001)
    Edited by D.L. Hawksworth

63. Brachiopods Past and Present (2001)
    Edited by H. Brunton, L.R.M. Cocks and S.L. Long

64. Morphology, Shape and Phylogeny (2002)
    Edited by N. MacLeod and P.L. Forey

65. Developmental Genetics and Plant Evolution (2002)
    Edited by Q.C.B. Cronk, R.M. Bateman and J.A. Hawkins

66. Telling the Evolutionary Time: Molecular Clocks and the Fossil Record (2003)
    Edited by P.C.J. Donoghue and M.P. Smith

67. Milestones in Systematics (2004)
    Edited by D.M. Williams and P.L. Forey

68. Organelles, Genomes and Eukaryote Phylogeny
    Edited by R.P. Hirt and D.S. Horner

69. Neotropical Savannas and Dry Forests: Plant Diversity, Biogeography,
    and Conservation
    Edited by R.T. Pennington, G.P. Lewis and J.A. Ratter

*Published by Academic Press for the Systematics Association
†Published by the Palaeontological Association in conjunction with Systematics Association
‡Published by the Oxford University Press for the Systematics Association
**Published by Chapman & Hall for the Systematics Association

T - #0134 - 111024 - C236 - 234/156/11 - PB - 9780367389987 - Gloss Lamination